Basic Mathematics
for Biochemists

8 95
.50

To Joy and Jenny

Basic Mathematics for Biochemists

ATHEL CORNISH-BOWDEN

Lecturer in Biochemistry,
University of Birmingham

LONDON   NEW YORK
CHAPMAN AND HALL

First published 1981 by
Chapman and Hall Ltd
11 New Fetter Lane, London EC4P 4EE
Published in the USA by
Chapman and Hall
in association with Methuen, Inc.
733 Third Avenue, New York NY 10017

Printed in Great Britain at the
University Press, Cambridge

ISBN 0 412 23000 3 (cased)
ISBN 0 412 23010 0 (paperback)

British Library Cataloguing in Publication Data

Cornish-Bowden, Athel
    Basic mathematics for biochemists.
    1. Biological chemistry   2. Mathematics
    I. Title
    510'.2454   QD42

    ISBN 0–412–23000–3
    ISBN 0–412–23010–0 Pbk

Contents

Preface

Some teachers of biochemistry think it positively beneficial for students to struggle with difficult mathematics. I do not number myself among these people, although I have derived much personal pleasure from the study of mathematics and from applying it to problems that interest me in biochemistry. On the contrary, I think that students choose courses in biochemistry out of interest in biochemistry and that they should not be encumbered with more mathematics than is absolutely required for a proper understanding of biochemistry. This of course includes physical chemistry, because a biochemist ignorant of physical chemistry is no biochemist. I have been guided by these beliefs in writing this book. I have laid heavy emphasis on those topics, such as the use of logarithms, that play an important role in biochemistry and often cause problems in teaching; I have ignored others, such as trigonometry, that one can manage without. The proper treatment of statistics has been more difficult to decide. Although it clearly plays an important part in all experimental sciences, it is usually preferable to treat it as a subject in its own right and not to try to incorporate it into a course of elementary mathematics. In this book, therefore, I have used a few examples from statistics to illustrate more general points, but I have not discussed it for its own sake. To summarize, the book is directed primarily towards students taking compulsory courses in mathematics in the early stages of their training as biochemists, but I hope it will also prove useful as a short revision text at later stages in the study of biochemistry.

I should like to thank my wife Mary Ann for her encouragement and support during the writing of this book, and for pointing out various ways in which it could be made more comprehensible. I am also grateful to my colleagues Geoffrey Bray, Stuart Ferguson, Baz Jackson, John Teale and Chris Wharton for reading the manuscript and suggesting many improvements to it.

January, 1981   Athel Cornish-Bowden

1 The Language of Mathematics

1.1 Introduction

Lord Kelvin once remarked that all of science could be divided into physics and stamp collecting. Although this rather patronizing comment of a physicist is not one that will appeal to all biochemists, it has an element of truth in it. A science can hardly claim to be a science as long as it remains no more than a catalogue of unrelated observations. Only when general laws can be proposed and tested by experiment can it be said to have passed from mere description into science. In chemistry the transformation from stamp collecting into science corresponded with the development of thermodynamics and, later, the atomic theory and theories of chemical bonding; in biochemistry, the gradual realization that understanding of life processes requires a foundation of physical chemistry and not just a list of metabolic reactions has played a corresponding role. It is no coincidence that mathematics has been central in all of these developments, and it is now almost impossible to comprehend even elementary biochemistry without a grasp of elementary mathematics. Fortunately for non-mathematically minded biochemists, however, the mathematics necessary for an undergraduate course in biochemistry *is* nearly all elementary, and nearly all of it has been touched on in every science student's previous education. Little more is required, therefore, than to identify the parts of elementary mathematics that are important in biochemistry and to reinforce them with appropriate examples.

In mathematics itself, the transformation from description into science is paralleled by the development from *arithmetic*, which is concerned with numbers and their manipulation, into *algebra*. Arithmetic is very useful, but it is much too limited to satisfy all of the needs of science. In arithmetic, every problem is a new and separate problem, and it is difficult to make useful generalizations and hence to express scientific laws. Let us consider the simple biochemical example in Table 1.1, which shows a set of rates of a reaction measured at the substrate concentrations given. As it stands the table is no more than a *description* of the results of a particular set of measurements and as long as we treat the numbers just as numbers we cannot make a general or useful statement about the enzyme to which they refer. The table tells us the rates observed at substrate concentrations of 5 mM and 10 mM but offers no guidance about what rate to expect at 8 mM; it does not tell us whether the system studied was behaving in accordance with some general law; it offers no clue as to what general law there might be. To remedy these omissions we must move beyond arithmetic into algebra, because only then will we be able to recognize a pattern or regularity in the numbers, and express it so that it can be recognized again if it occurs with another system. If the rates in Table 1.1 are represented as v and the substrate concentrations as s, then the following equation expresses a *law* that defines all of the numbers in the table:

$$v = \frac{10s}{4+s}$$

This equation has two advantages over the table: first, it allows a *summary* of all of the information while occupying much less space; secondly, it *predicts* what v values we might expect to observe at s values that are not included in the table. For example, it answers the question above by predicting that $v = 6.67 \ \mu M\,s^{-1}$ when $s = 8$ mM. This is clearly more useful than just listing a set of numbers recorded on a particular occasion.

One can proceed one stage further with this example by noting that the equation is typical of what is reported for many enzymes, and so if

Table 1.1 Observations from a kinetic experiment

Substrate concentration (mM)	Rate ($\mu M\,s^{-1}$)
1	2.00
2	3.33
5	5.56
10	7.14
20	8.33

we replace the numbers 10 (μM s^{-1}) and 4 (mM) by V and K_m, respectively, we have an equation that expresses a *generalization* about enzymes:

$$v = \frac{Vs}{K_m + s}$$

Again, replacing numbers with *symbols* has increased the generality of what we want to say. In addition to the symbols v, V, s and K_m, which represent numbers, either particular ones or generalized ones, the equation contains three *operators*: one is represented by the addition sign $+$; one is shown by the horizontal line between Vs and $K_m + s$; and the third is implied by the juxtaposition of V and s, but could have been made explicit by writing $V \cdot s$ instead of Vs. Each operator specifies something to be done to the numbers or symbols operated on: the $+$ sign requires K_m and s to be added together; the horizontal line requires Vs to be divided by $K_m + s$; the juxtaposition of V and s (or a dot between them) requires them to be multiplied together.

As long as no ambiguity is possible mere juxtaposition is sufficient to indicate multiplication, but if several numbers are to be multiplied together, or if we allow algebraic symbols consisting of more than one letter each (as in Fortran and many other computer languages), or if ambiguity is possible for some other reason, multiplication can be indicated by a dot or a cross. The dot is more common for pairs of symbols (where no confusion with the decimal point is possible) and the cross is more common for pairs of numbers, but the two symbols have the same meaning in most contexts, i.e. $V \times s$ means the same as $V \cdot s$. (In some specialized applications, such as in *vector algebra*, it is convenient to assign distinct meanings to $\cdot$ and $\times$, but these need not concern us in elementary biochemistry). In current usage the dot should be written *above* the line and the decimal point *on* the line, e.g. $5.1 \cdot 8.7 = 44.37$ *not* $5 \cdot 5 \cdot 8 \cdot 7 = 44 \cdot 37$ but this is rather a recent convention so far as British books are concerned: in older British (but not American) work one is likely to encounter exactly the opposite convention.

1.2 Priority rules for operators

To avoid ambiguity it is important to realize that operators have to be obeyed in a proper order. Unlike ordinary language, equations are not read from left to right but in accordance with *priority rules* that require certain operators to be obeyed before others. Thus, the value of $5 \times 3 + 2 \times 4 - 3$ is 20, not 25 or 65, because multiplication must be done before addition or subtraction. In general, the rule is as follows:

(1) expressions within brackets must be evaluated first;
(2) if brackets are 'nested' (brackets within brackets), 'inner' brackets must be evaluated before 'outer';
(3) exponentiation must be carried out before multiplication and division;
(4) multiplication and division must be carried out before addition and subtraction.

'Exponentiation' is the raising of a number or expression to a power, as in x^2, $(x + y)^a$, etc. If a power is itself raised to a power we work down from the top, i.e.

$$e^{-2x^2} \text{ means } e^{(-2x^2)} \text{ } not \text{ } (e^{-2x})^2$$

As in this example, it is always permissible and often desirable to use brackets to clarify an expression that might otherwise be misinterpreted. This is true even if the expression without brackets is strictly unambiguous.

There are no rules of priority between addition and subtraction among themselves, because the result of a sequence of additions and subtractions is independent of the order in which they are done. This is normally a matter of convenience only, although occasionally numerical considerations may make one order better than another. In principle, the same applies to multiplication and division, but greater care is needed because thoughtless use of the slash / to indicate division often results in expressions with meanings that are either unclear or clear but different from what the writer intended. It is wisest therefore to use the slash in moderation and to check carefully that expressions have the meanings intended. Consider for example the following equation:

$$v = \frac{Vs}{K_{\mathrm{m}}(1 + i/K_{\mathrm{i}}) + s}$$

This is unambiguous, and the priority rule should prevent the bracketed expression $(1 + i/K_{\mathrm{i}})$ from being misread as $[(1 + i)/K_{\mathrm{i}}]$. When there is more than a single term after the slash, however, as in $(i/K_{\mathrm{i}} + 1)$, misunderstanding is more likely because it is not always clear whether the slash indicates simple division or whether it is used to avoid the typographical inconvenience of a cumbersome fraction such as $\dfrac{i}{K_i + 1}$. Double slashes are so confusing that they should *never be used*: this applies not only to ordinary algebraic expressions but also the units of physical quantities, as in $R = 8.314$ J/mol/K. Here it is not clear whether the K belongs in the numerator of the unit with the J or in the denominator with the mol. To avoid this uncertainty the definition should be written as follows: $R = 8.314$ J mol^{-1} K^{-1}. For reasons that will become clear in Chapter 2, mol^{-1} and K^{-1} have the meanings (1/mol) and (1/K), respectively. In general, slashes should only be used in expressing units when there is only a single term in the denominator.

Certain computer languages do not obey precisely the same priority rules as conventional mathematics. This fact has generated rather more confusion about the conventions than existed before computers became widespread, and the appearance of cheap electronic calculators has made matters much worse in this regard because those that use so-called 'algebraic notation' commonly ignore mathematical conventions altogether and use a 'left-to-right' system. An expression such as

$$A * A/B/C/D + A/B * C$$

would be unambiguous in a Fortran program and would have the meaning

$$\frac{AA}{BCD} + \frac{AC}{B}$$

(The multiplication sign in Fortran is expressed as * and must be explicit). This unambiguous meaning, which may nonetheless be different from what the programmer intended, does *not* imply that such expressions are acceptable in ordinary mathematics. Similarly, the fact that simple calculators often disregard priority rules does not mean that they are obsolete. For example, the expression $5 \times 3 + 2 \times 4 - 3$ must be interpreted as $(5 \times 3) + (2 \times 4) - 3$ if the proper conventions are followed, even though most simple calculators using so-called 'algebraic notation' execute instructions as they are entered and consequently interpret the above expression as $\{[(5 \times 3) + 2] \times 4\} - 3 = 65$ (this is indeed what I get if I key in $5 \times 3 + 2 \times 4 - 3 =$ on my pocket calculator).

1.3 The summation sign

It often happens, especially in statistical calculations, that we need to add together a large number of terms of the same kind. For example, if we have a set of values $x_1, x_2, x_3 \ldots x_n$, their arithmetic mean $\bar{x}$ is given by

$$\bar{x} = (x_1 + x_2 + x_3 + \ldots + x_n)/n$$

In more complex examples explicit representation of the summation becomes cumbersome and unnecessary, and it is more convenient to use a special operator called the *summation sign* $\sum$ (a capital Greek sigma) instead:

$$\sum_{i=1}^{n} x_i \equiv x_1 + x_2 + x_3 + \ldots + x_n$$

The *limits* $i = 1$ and n written above and below the sign mean 'start adding at $i = 1$ and continue until $i = n$'. If the limits are obvious, as for example in statistical calculations where one often has to sum over all of a set of observations, they can be omitted.

Although the summation sign is a considerable convenience when the underlying calculation is understood, it can sometimes obscure the meaning when it is not. Indeed, one of the main reasons why more advanced mathematics than one is familiar with can appear much more difficult than it actually is, is that it often uses special notation to express results more compactly. For example, the whole of *matrix algebra* is a way of expressing very complicated relationships in an extremely compact way: very convenient when one is familiar with the symbolism but baffling when one is not. Whenever obscurity threatens for this sort of reason it is often helpful to translate the compact expressions into a more long-winded form, and then their meanings are likely to become clearer.

A corresponding sign for *products* also exists, although it is much less often encountered than the summation sign. It is written as the Greek capital pi, $\prod$, and is used in an exactly analogous way, i.e.

$$\prod_{i=1}^{n} x_i \equiv x_1 x_2 x_3 \ldots x_n$$

1.4 Functions

A mathematical *function* can be regarded as a set of instructions to carry out a series of operations on a variable or set of variables. For example, if we define v in terms of s as

$$v = \frac{Vs}{K_m + s}$$

where V and K_m are constants, then we are defining v as a *function* of s, by defining what operations have to be carried out on s to obtain v. We can also have functions of more than one variable. For example, v may be determined not solely by a single concentraion s but may depend both on s and on another concentration i:

$$v = \frac{Vs}{K_m(1 + i/K_i) + s}$$

and now we say that v is a function of both s and i.

Sometimes we may wish to symbolize the existence of a dependence of one variable on another without specifying what the dependence is. We then often use the symbol $f(\ )$ or something similar, e.g.

$$v = f(s, i)$$

which states that v depends on s and i but does not indicate whether the dependence follows the equation given above or some other equation.

There are a number of functions that are so often required in mathematics that they are given special symbols. For example, if y is always given by multiplying x by itself we say that y is the *square* of x and symbolize the relationship as

$$y = x^2$$

where the superscript 2 indicates that 2 x's need to be multiplied together. Conversely, x in this example is the *square root* of y, which we may write as

$$x = \sqrt{y}$$

or, more commonly and for reasons that I shall discuss in Chapter 2, we may express the same relationship as

$$x = y^{\frac{1}{2}}$$

Other functions of great importance are the *logarithmic* and *exponential* functions, which I shall also consider in Chapter 2, the *derivative*, or result of differentiation (Chapter 3), and the *integral* (Chapter 4).

There are others, such as *trigonometric functions*, that are important in mathematics generally, but have little application in elementary biochemistry, and so I shall say little about them. On the other hand, there are certain functions that have little importance in mathematics as a whole but which are useful to define for biochemical purposes. For example, various properties of proteins can be related to the hydrogen-ion concentration $[H^+]$ in terms of the following kind of expression:

$$y = \frac{\tilde{y}}{1 + ([H^+]/K_1) + (K_2/[H^+])}$$

in which $\tilde{y}$, K_1 and K_2 are constants. This kind of function was first studied by Michaelis and it is consequently called a *Michaelis function*.

1.5 Constants, variables and parameters

Some quantities, such as the number 2.0, have a unique value under all circumstances and are called *constants*. Other numbers, such as the *gas constant* $R \simeq 8.3 \, J \, mol^{-1} \, K^{-1}$ are found by experiment to be constant also. Others, such as K_m in the Michaelis–Menten equation,

$$v = Vs/(K_m + s)$$

may be constant for a particular enzyme and substrate under well-defined and constant conditions, although they may vary with temperature, pH, etc. These quantities can be treated mathematically as constants only as long as the physical conditions that determine them are constant.

We are often interested in quantities that change when the conditions change. For example, we may find that an equilibrium 'constant' K varies with the temperature T according to the van't Hoff equation:

$$K = \exp\left(\frac{\Delta S^0}{R} - \frac{\Delta H^0}{RT}\right)$$

where $\exp(\,)$ is the exponential function (Chapter 2), and ΔS^0, ΔH^0 and R are constants. Thus although K may be constant at constant T it varies with T and so if we are concerned with changes in temperature we must treat K as a *variable*.

It will be evident that the distinction between a constant and a variable is not absolute, because any variable becomes a constant if the conditions that determine its variation are kept constant. In some contexts the distinction is so difficult or inconvenient to make that we introduce a third term, *parameter*, to denote a quantity that is treated as variable for some purposes and constant for others. This occurs especially when we use statistical methods to estimate unknown constants: although we may believe it to represent a true physical constant we must treat it as a variable during the process of estimation. Suppose, for example, that we have a measured quantity y that depends on a controlled variable x according to the following linear equation:

$$y = a + bx + \varepsilon$$

in which a and b are the physical constants whose values we require and the third term ε on the right-hand side represents *experimental error*. This error term prevents us from calculating a and b exactly from measurements of y at two values of x. Instead, we would measure y at *several* values of x and try to find values of a and b that predicted the observed y values as closely as possible (I shall discuss a similar problem in more detail in Chapter 6). During the analysis we are free to try any values of a and b that we like: in other words, we treat them as variables, even though we may believe that some physical reality exists in which they have constant, but unknown, values. To express this duality we use the word *parameter* for a quantity such as a or b that controls the dependence between our observable variables.

1.6 Dimensional analysis

Most of the quantities we measure in biochemistry are not simple numbers but numbers with *units*. For example, the concentration of glucose in blood is not 0.005 but 0.005 mol dm^{-3}. Put differently, and very formally, we may say that the blood–glucose concentration has the *dimensions* of (amount of substance) (length)$^{-3}$. Actually, this combination of dimensions occurs so often in biochemistry that it is convenient to define a *unit* of *amount-of-substance concentration*, such that 1 molar or 1 M $\equiv$ 1 mol dm^{-3}; and because amount-of-substance concentration is the only kind of concentration the biochemist is usually interested in we usually abbreviate this cumbersome term to *concentration*. Although strictly it is a derived quantity it is often convenient to treat it as if it were a primary dimension.

Consideration of units and dimensions is sometimes regarded as a pedantic nuisance, but this is a pity, because it is one of the most powerful tools that scientists have for detecting mistakes in algebra – not only their own but also other people's. This is because there are rules that govern the way in which dimensioned quantities can be combined and a high proportion of algebraic mistakes cause these rules to be violated.

The simplest dimensional rule is that one cannot equate quantities of different dimensions: it is meaningless to assert that a length of 3 cm is equal to a mass of 1 kg, for example. This may seem so obvious as to be hardly worth mentioning, but it is surprising how many fallacies in kinetics spring from an inability to appreciate that one cannot compare a first-order rate constant with a second-order rate constant, because their dimensions are different. An obvious extension of the first rule is that one cannot add or subtract quantities of different dimensions or say that they are greater or less than one another. On the other hand we can multiply them together or divide one by the other: if an object has a mass of 1 g and a volume of 2 cm^3 it is quite acceptable to divide 1 g by 2 cm^3 to obtain a density of 0.5 g cm^{-3}.

For certain mathematical purposes only *pure numbers*, i.e. numbers with no dimensions, are admissible. For example, the expression 2^i means that i 2's have to be multiplied together, and this has meaning only if i is a pure number: we cannot multiply 3 cm 2's together, for example. In other words, dimensioned quantities must not appear as *exponents*, and, similarly, their *logarithms* cannot be taken.

Sometimes we may wish to define a quantity, such as pH, that appears to be the logarithm of a dimensioned quantity, namely the hydrogen-ion concentration in the case of pH. However, as I discuss in Chapter 2 (p. 37), we can only do this if we first remove the dimensions by dividing by a *standard value* that has the same dimensions.

The value of all this is that when we make a mistake in algebra we often introduce a dimensionally incorrect expression. So, if we arrive at a result that we suspect may be incorrect, the simplest check we can make is for dimensional consistency. Suppose, for example we are trying to remember the *Henderson–Hasselbalch equation* (Chapter 3), and think that it may be

$$pH = pK_a + \log\{[salt][acid]\}$$

Is this likely? We can check by noting that [salt] and [acid] are both concentrations and therefore their product is a concentration squared. As this is not a pure number it cannot have a logarithm and so the expression is wrong. If we are reasonably sure we have the right ingredients but that we have combined them incorrectly, we can use knowledge of the dimensions to deduce that we need a ratio of concentrations, which would be a pure number, not a product. But do we want [salt]/[acid] or [acid]/[salt]? Here dimensional analysis cannot help us, but our knowledge of chemistry can: we should know that the pH decreases as a solution becomes more acid, and this should guide us to the correct form of the equation:

$$pH = pK_a + \log\frac{[salt]}{[acid]}$$

If we have a long derivation that begins dimensionally correct but ends at a dimensionally incorrect result, we know that it must contain a mistake. How can we find it? One way might be to repeat the derivation, but then we may well repeat whatever thought processes led to the original mistake, in which case we shall not find it. A better approach – better because it is simpler, quicker and requires *new* thoughts rather than a repetition of old ones – is to check the dimensions at various stages until the mistake is located.

Dimensional analysis is also useful for remembering the slopes and intercepts of graphs. Anything measured along the x-axis, such as the intercept of a line on the x-axis, must have the same dimensions as x; anything measured along the y-axis must have the same dimensions as y. Furthermore, the *slope* at any point must have the dimensions of y divided by those of x. The reason for this will become clear in Chapter 3 when we consider the definition of a slope as the limit of a change in the y co-ordinate divided by the corresponding change in the x co-ordinate as one moves along the line. For the present we can accept it as a fact without worrying about the reason.

The rules for applying dimensional analysis to graphs are illustrated in Fig. 1.1. They allow us to check whether we have remembered the slopes and intercepts of particular plots correctly, and can guide us to the correct ones when we have not.

Some authorities advocate converting all variables into dimensionless form before plotting them. For example, instead of plotting [S], a

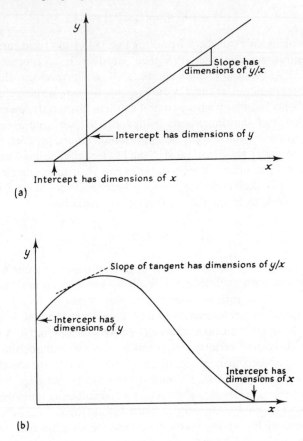

Fig. 1.1. Dimensional analysis as applied to graphs: (a) straight line; (b) arbitrary curve.

concentration, in mM, we would make it dimensionless by dividing it by a 'standard' concentration of 1 mM, so that we actually plot $[S]/mM$, a dimensionless variable. Similarly, one can avoid the problems of defining thermodynamic quantities in terms of the logarithms of dimensioned equilibrium constants by defining all equilibrium constants in such a way as to make them dimensionless. My own view is that the advantages of following these recommendations are very slight and the disadvantages are considerable. If all variables are dimensionless dimensional analysis becomes meaningless and one denies oneself the use of a valuable tool. One can also very easily lead oneself into absurdities: by insisting tht equilibrium constants are dimensionless, for example, one prevents the Michaelis constant K_m from being compared with the dissociation constant of the enzyme–substrate complex unless one defines K_m so that it is dimensionless as well; but doing this causes the Michaelis–Menten equation to contain a dimensionally incorrect addition of K_m to the substrate concentration – unless of course one is willing to have a dimensionless substrate concentration too! One recent and widely used book of problems in physical chemistry for biochemists falls into this particular trap.

1.7 Plotting graphs

Biochemistry is not a visual subject. Unlike zoology and botany, and to some extent even such mathematically based subjects as physics and astronomy, it rarely presents the experimenter with much to see directly. Nearly always we have to 'see' the properties of a biochemical system by way of numbers measured on an instrument such as a spectrophotometer. The first stage in translating these rather abstract observations into biochemical knowledge is often the drawing of a *graph* to show how the observed variable depends on one or more controlled variables. It is therefore vitally important for the biochemist – more than almost any other kind of scientist – to be familiar with the rudiments of plotting graphs so that they display the information they contain in the clearest way.

Rather than spend a large amount of text discussing the ways of plotting data well and badly I have chosen a number of examples to illustrate common faults in plotting. These are shown in Fig. 1.2. In (a), some observations are plotted with very poorly chosen scales on both axes. Less than a fifth of the ordinate scale is actually used, with the result that most of the graph is uninformative white paper. In (b), the ordinate scale is much better, but because of the large gap in data between $x = 10$ and 50 most of the observations are crowded together on the left and there is still too much empty space. In fact most of the information is in these low-x data points and the one at $x = 50$ was probably measured to give an idea of the limit approached by y. There is no real need to show the region between $x = 10$ and 50. We can then choose a scale that will spread out the low-x points, but break it to include the $x = 50$ point, as in (c). Whenever the scale is broken in this

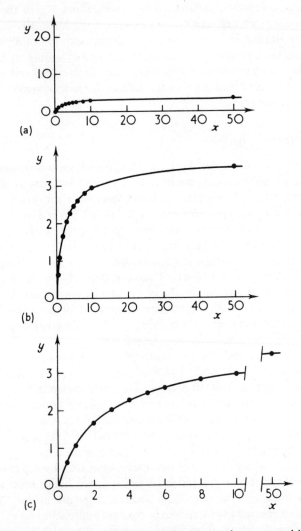

Fig. 1.2.   Common faults in plotting graphs: (a) wasted space caused by poorly chosen scales; (b) crowding caused by poor choice of abscissa scale; (c) satisfactory plot of the same data;

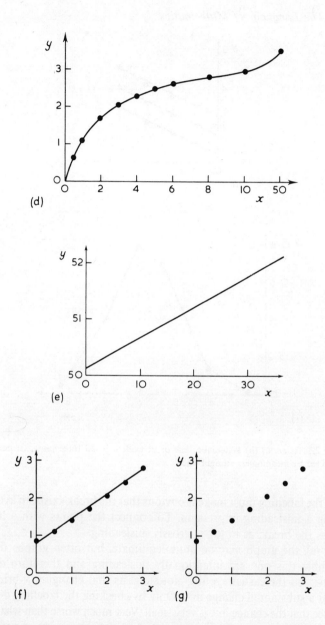

Fig. 1.2.   (d) spurious inflexion caused by failing to note a sudden change in scale; (e) misleading axes suggest a much larger slope than actually exists; (f) straight line drawn through points that demand a curve; (g) same data without the line;

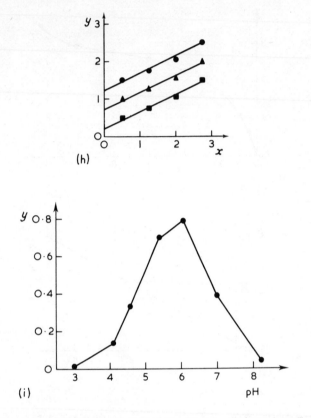

Fig. 1.2.   (*Contd.*) (h) systematic lack of fit evident in all three plots; (i) points connected by meaningless straight lines.

way, the labelling must make it obvious that the break exists, to avoid giving a misleading impression. To connect the points with a line across the break, as in (d), is grossly misleading.

In (e), the graph *may* be quite legitimate, but often graphs that resemble this one are intentionally misleading and therefore dishonest. To the casual eye the plot suggests that changing x brings about a substantial change in y. Only by checking the labelling does one see that the change in y is very small. Very much worse than what is shown here is to present the same sort of graph but to omit some or all of the labelling. Politicians and advertisers are fond of this kind of graph but it has no place in science.

The graph in (f) shows a very common fault: a straight line is drawn through points that demand a curve. The same points are shown in (g) without the line, and the curvature is somewhat more obvious: a line tends to bias the eye and it is prudent, therefore, to examine a set of points carefully before drawing a line through them. It is illuminating to take any issue of any research journal of biochemistry at random and search for examples of straight lines drawn through points that do not fit straight lines. They are not hard to find. A useful habit to cultivate is to examine all experimental graphs – your own and other people's – with your eye close to the paper and looking along the plotted line (Fig. 1.3). This greatly emphasizes any systematic failure of the points to lie on the line. In real experiments, of course, there is always some experimental error, so we should not expect an exact fit. But experimental error should be random, and so experimental points should be randomly scattered about the line, not systematically.

When there are very few points on each line, as in (h), it is very difficult to distinguish between random scatter and systematic lack of fit. One answer is of course to obtain more observations, but if several lines show the same sort of lack of fit, as in (h), there is good reason to believe it is systematic.

Finally, (i) shows a fault that seems to be particularly common in plots of pH dependences. Although any reasonable model would place the points on a smooth curve, they have been connected by

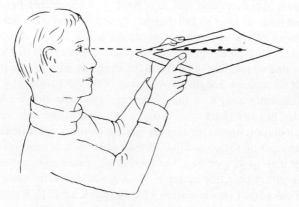

Fig. 1.3.   How to inspect a graph for lack of fit.

straight-line segments. This is not only unattractive, it also misrepresents the form of the dependence, especially in regions of high curvature: in (i) the maximum ought to be at pH 5.85, not at pH 6.1, and it should be at a height of 0.82, not 0.79. In experiments where there has been no attempt to account for the observations mathematically it may sometimes be justifiable to connect points by straight-line segments simply to display them more prominently, but this should never be done in work with any mathematical pretensions.

1.8 Precision

One of the commonest faults in scientific writing is to report results with an inappropriate degree of precision. For example, if a sample of 100 ml of blood is found to contain 93 mg of sugar, the concentration in mM (expressed in terms of glucose, of relative molecular mass 180) is $1000 \times 93/(100 \times 180)$, which can be variously written as 5 mM (to one *significant figure*), or 5.17 mM (to three significant figures), or even 5.166 666 7 mM (to 8 significant figures), etc. Which of these is appropriate for reporting the result? Common sense tells us that 5 mM is rather more vague than the data would permit, whereas 5.166 666 7 mM is ridiculously precise, and 5.17 mM seems a reasonable compromise; but how can we justify such a compromise logically? If we examine the four numbers used in the calculation, we see that 1000, the factor for converting from M to mM, is exact and we could write it as 1000.000 . . . if we wished, 93 mg and 100 ml are of unknown accuracy, and 180 is correct to 3 significant figures, i.e. the true value is closer to 180 than to 179 or 181. Without knowing the conditions of the experiment we cannot know how accurately the mass and volume were measured, but it seems reasonable to guess that the volume was chosen to be 100 ml by the experimenter and that it is correct to at least 3 significant figures. This leaves the 93 g as the probable main source of inaccuracy and the main one we have to consider: just as a chain is no stronger than its weakest link, the result of a calculation (in simple cases at least) is no more accurate than the most inaccurate value used in it. Presumably the true mass was closer to 93 g than to 92 g or 94 g, but there is no reason to suppose it was 93.00 g rather than, for example, 92.68 g, but this latter value would give a calculated concentration of 5.148 888 9 mM (to 8 significant figures). Comparing this with the original value of 5.166 666 7 mM we see that only the first two digits are reliable, though the third is not worthless: consequently 5.17 mM seems a reasonable way of expressing the concentration, with the understanding that the last (or *least significant*) digit is unreliable. To include any more than three significant figures in this result would be to claim that it is more accurate than it is. As a general rule one should express a result with no more (or only a little more) precision than the accuracy of the data from which it was calculated.

Including too many significant figures may seem to be a harmless fault – after all, the reader can always ignore the unwanted digits. It is not harmless, however, because it can mislead the reader into thinking you have measured more than you have, and it can make you look a fool. Expressing the concentration calculated above as 5.166 666 7 mM is just as silly as claiming that the fact that grass is green proves that it contains chlorophyll. The observation and the conclusion in this case are both true, and they are relevant to one another; nonetheless, the conclusion does not follow from the observation. Similarly, the true concentration of glucose could conceivably be 5.166 666 7 mM, but the available information does not justify saying so.

Inaccuracy can result not only from inaccuracy in the data; it can also arise from imprecise arithmetic during the calculation, especially if several steps are involved. It is advisable therefore to carry out the intermediate stages of a calculation with more precision (more significant figures) than one expects to retain at the end. However, it is not only very tedious to carry out every step with, for example, fifteen significant figures if only three are to be retained at the end; it is also unnecessary and it encourages mistakes. A fairly reliable rule is to use *one more significant figure* in the calculations than one expects to retain in the final result. This rule needs to be broken, however, in calculations that involve small differences between nearly equal numbers: for example, if a beaker weighs 10.381 4 g empty but 10.493 1 g after addition of a sample of a chemical it would be unwise to calculate the weight of chemical by substracting 10.4 g from 10.5 g because this would give only one significant figure in the answer whereas the data would support four.

At the end of a calculation the unwanted digits have to be discarded. This can be done either by *truncation* or by *rounding*; thus 9.358 1 can be truncated to 9.35 or it can be rounded to 9.36. In truncation we simply omit the unwanted digits; in rounding we increase the last retained digit by 1 if the portion of the number to be discarded begins with 5, 6, 7, 8 or 9. Rounding is slightly more accurate than truncation, and is the usual practice in science.

In our own measurements we know (or should know) how precise they are, but in other people's there may be ambiguity. We had an example of this at the beginning of this section: the value of 180 given for the relative molecular mass of glucose is actually correct to three significant figures, but there is no way of knowing this from the way the number is written; it would still be written as 180 if the true value were 177 but for some reason was given to only two significant figures. If it is important to avoid this ambiguity we can state the precision explicitly, for example by giving the value as '180 (correct to three significant figures)', but this is cumbersome for ordinary use and we normally assume that we can gauge the precision by counting the number of significant figures. This is somewhat more subtle than it sounds, because although all non-zero digits are counted, zeroes may or may not be counted depending where in the number they occur. The simplest way of counting significant figures (which is easier to do than to describe) is to proceed as follows: (1) identify the first significant digit, which is the first non-zero digit encountered on reading the number from left to right; (2) identify the last significant digit, which is the last non-zero digit if the decimal point is not explicitly included, *but* the last digit of any kind if the decimal point is included; and (3) count all the digits (including zeroes), starting from the first and finishing with the last significant digit. This procedure may be illustrated by the following examples, in which significant zeroes are shown in italics and the number of significant figures is shown in parenthesis after each number: 107 (3); 13.064 (5); 0.120 (3); 0.12 (2); 100 (1); 100.0 (4); 003 (1). (Zeroes are not usually written in front of integers, as in 003, which would normally be written simply as 3, but if they are written they are not significant.)

We do not usually italicize significant zeroes and it is cumbersome to add a parenthesis specifying the precision. There is, however, a third way of indicating that a value such as 100 ml is intended to have three significant figures: we can choose different units that make the precision clear, e.g. we can write it as 0.100 litre, or we can achieve the same result by relating the value to a power of 10 (see Section 2.1, p. 25), e.g. by writing it as 1.00×10^2 ml. These representations are often used in scientific writing.

Statistical methods are not discussed in this book, but a mention of them is appropriate here because they can provide estimates of the precision of calculated values with less of the guesswork and plausibility arguments that have characterized much of this section. When such a precision estimate exists it can be written after the value prefixed by a $\pm$ sign ('plus-or-minus') thus: 5.17 ± 0.08 mM. There is rarely any point in reporting a precision estimate with more than one or two significant figures, and the value it qualifies should be reported with the same number of *decimal places*, i.e. the same number of significant digits after the decimal point. If there is no decimal point the two numbers should have the same number of insignificant zeroes. Thus we might write 81.031 ± 0.006 or $135\,200 \pm 1400$, but there would be little point in writing $135\,217 \pm 1400$ because the right-hand 17 would clearly be meaningless and so better expressed as 00.

Finally, I should say a word about the difference between *accuracy*, which is concerned with the truth of what we say, and *precision*, which is concerned with how we say it. For example, the constant π has a value of about 3.141 59, and so the expression $\pi = 3.589\,37$ may be very precise but it is not at all accurate, whereas the expression $\pi = 3.14$ is much less precise but much more accurate. Precision is always within our control, but accuracy often is not. The main theme of this section has been that we should aim to express results so that their precision is only a little greater than their accuracy: less precision than this discards good information; more makes us look foolish.

1.9 Problems

(1.1) If $x_1 = 1$, $x_2 = 3$, $x_3 = 2$, $x_4 = 7$, $x_5 = 6$, $x_6 = 9$, what are the values of the following expressions?

(a) $\displaystyle\sum_{i=1}^{6} x_i$; (b) $\displaystyle\sum_{i=2}^{3} x_i$; (c) $\displaystyle\sum_{i=1}^{4} i x_i$; (d) $\displaystyle\sum_{i=1}^{6} x_i^2$

(1.2) The *sample variance* s^2 of a set of n numbers, $x_1, x_2 \ldots x_n$, can be defined as

$$s^2 = \frac{1}{n} \sum_{i=1}^{n} (x_i - \bar{x})^2, \text{ where } \bar{x} = \frac{1}{n} \sum_{i=1}^{n} x_i$$

Show that this is equivalent to

$$s^2 = \frac{1}{n} \sum_{i=1}^{n} x_i^2 - \frac{1}{n^2} \left(\sum_{i=1}^{n} x_i \right)^2$$

(1.3) The Nernst equation asserts that

$$E = E^0 + \frac{RT}{n\mathscr{F}} \ln \frac{[\text{ox}]}{[\text{red}]}$$

where E and E^0 are potentials measured in volts, $R = 8.314 \, \text{J mol}^{-1} \, \text{K}^{-1}$ is the gas constant, T is the temperature in K, n is the number of electrons transferred in the oxidation-reduction process, $\mathscr{F}$ is a constant with the value 96 494 coulomb mol^{-1}, and $[\text{ox}]$ and $[\text{red}]$ are the concentrations of the oxidized and reduced forms, respectively, of the redox couple. Assuming that $1 \, \text{V} \equiv 1 \, \text{J A}^{-1} \, \text{s}^{-1}$ and 1 coulomb $\equiv 1 \, \text{A s}$, show that the equation is dimensionally consistent.

(1.4) Which of the following equations or statements must be incorrect because they are dimensionally inconsistent? ($[\text{S}]$, $[\text{I}]$, K_{m} and K_{i} are concentrations; v and V are rates, i.e. concentrations divided by time).

(a) $v = \dfrac{V[\text{S}]}{K_{\text{m}} + [\text{S}] + [\text{I}]/K_{\text{i}}}$

(b) The intercept on the $[\text{S}]/v$ axis of a plot of $[\text{S}]/v$ against $[\text{S}]$ is K_{m}/V

(c) The slope of a plot of v against $v/[\text{S}]$ is $-1/K_{\text{m}}$.

(1.5) Evaluate the following expressions:

(a) $8 + 6/3 + 4$
(b) $8 + 6/(3 + 4)$
(c) $(8 + 6)/(3 + 4)$
(d) $2^3 + 3^2$
(e) $2^3 \times 3^2$
(f) 2^{3^2}
(g) $(2^3)^2$

(1.6) Different methods of determining the average relative molecular mass M_{r} of macromolecules in solution yield different kinds of average. For example, osmotic-pressure measurements yield a *number average* M_{n}, light-scattering measurements yield a *weight average* M_{w}, and equilibrium sedimentation yields a *Z-average* M_{z}. These are defined as follows:

$$M_{\text{n}} = \frac{\sum n_i M_i}{\sum n_i}; \quad M_{\text{w}} = \frac{\sum n_i M_i^2}{\sum n_i M_i}; \quad M_{\text{z}} = \frac{\sum n_i M_i^3}{\sum n_i M_i^2}$$

where n_i is the number of mol of a species with $M_r = M_i$ and the summations are carried out over all solute species. Evaluate all three averages for a solution of protein existing as 27 nM monomer, 17 nM dimer and 1.3 nM tetramer, assuming that $M_r = 72\,000$ for the monomer.

(1.7) Give the results of the following calculations with appropriate precision:

    (a) $17.3 \times 1.382\,21$
    (b) $2.571/(6.331 - 2.111)$
    (c) $1.3359 \times (35.6579 - 35.6112)$
    (d) 2.1×4.361

(1.8) How many significant figures does each of the following values contain?

    (a) $23.007\,050$
    (b) $0.007\,050$
    (c) $135\,000$
    (d) 1.350×10^5
    (e) 10.37

2 Exponents and Logarithms

2.1 Integer powers

In mathematical manipulations one often has to multiply the same number by itself repeatedly, so that one has such expressions as

$$2 \times 2 \times 2 \times 2 \times 2 = 32$$

This sort of thing can rapidly become both tedious and difficult to read, and it is convenient to have a shorthand notation that conveys the same information not only more concisely but also (once one is used to it) more clearly. *Exponents* fulfil this need. The simplest use of an exponent *i* is to raise a number to its *i*th *power*, i.e. to show that the number is to be multiplied by itself *i* times, the exponent being written a little above the number that it operates on. For example,

2^2 means $2 \times 2 = 4$
2^3 means $2 \times 2 \times 2 = 8$
2^4 means $2 \times 2 \times 2 \times 2 = 16$
3^3 means $3 \times 3 \times 3 = 27$

and so on. An exponent can have the value 1, but then it is usually omitted because the number it operates on is unaltered: $2^1 = 2$, $3^1 = 3$, $4^1 = 4$, etc.

It might seem that an exponent would have to be a positive integer: if this were so exponents would be much less useful than they are. It is, however, possible to assign plausible meanings to zero, negative and fractional exponents and thereby to extend their usefulness very greatly. Consider the following pair of series:

$$i = 1 \quad 2 \quad 3 \quad 4 \quad 5 \quad 6 \quad \ldots$$
$$2^i = 2 \quad 4 \quad 8 \quad 16 \quad 32 \quad 64 \quad \ldots$$

What rules govern the choice of numbers to be put in each lines? Each value of i is greater by 1 than the value on its left, whereas each value of 2^i is double the value on its left. What happens if we read the series from right to left? The rules can simply be reversed: each value of i is 1 *less* than the value on its right, and each value of 2^i is *half* the value on its right. There is nothing in these rules, however, that requires us to stop at $i = 1$, and in fact we can continue indefinitely (reading from right to left):

$$\ldots \quad -4 \quad -3 \quad -2 \quad -1 \quad 0 \quad 1 \quad 2 = i$$

$$\ldots \quad \frac{1}{16} \quad \frac{1}{8} \quad \frac{1}{4} \quad \frac{1}{2} \quad 1 \quad 2 \quad 4 = 2^i$$

We get similar results for any number, not just 2, raised to the ith power, and in general we can define:

$$a^3 = aaa$$
$$a^2 = aa$$
$$a^1 = a$$
$$a^0 = 1$$
$$a^{-1} = 1/a$$
$$a^{-2} = 1/a^2$$
$$a^{-i} = 1/a^i$$

The relationship $a^0 = 1$ must be noted particularly. Although it may seem surprising at first sight it arises quite naturally from the series that was discussed above.

The main use of integer powers greater than the second or third in biochemistry is for convenient representation of very large or very small numbers. For example, the number of molecules in 1 mol of a substance is about 600 000 000 000 000 000 000 000, whereas the number of gram-ions of H^+ in a litre of 1M–NaOH is about 0.000 000 000 000 01. Numbers such as these are very inconvenient to manipulate and even to read if they are written in the ordinary way. (This is perhaps borne out by the fact that in typing the manuscript of this book I needed three attempts at typing this last number before it was correct.) In practice, therefore, it is usual to relate very large or very small numbers to the nearest power of 10: thus, the first of the two numbers mentioned is 6 times greater than 10^{23}, and is therefore usually written as 6.0×10^{23}, whereas the second is equal to 10^{-14}. In practice in experimental science we usually simplify even further, at least with dimensioned quantities, by using multiples or submultiples of units: for example, the biochemist often has to deal with concentrations around 0.000 01 M. This could be written as 10^{-5} M, but it leads to fewer mistakes in arithmetic if we use a smaller unit of concentration and write it as 10 μM, where the μ (spoken aloud as 'micro' in this use) is a standard prefix meaning 10^{-6}.

In principle we could relate large and small numbers in this way to powers of any number, but it is normally convenient to use powers of 10 because 10 happens to be the basis of our ordinary number system. To avoid confusion we always use powers of 10 even for numbers that have simple expressions as powers of other numbers. Thus we would express 2401 as 2.401×10^3 even though it happens to be precisely 7^4.

2.2 Fractional exponents

We have seen that the effect of increasing the exponent in an expression a^i by 1 is to cause the value of the expression to be multiplied by a; we shall now consider the effect of doubling the exponent. In this case we produce the *square* of a^i, i.e. $a^{2i} = (a^i)^2$. For example: $2^2 = 4, 2^4 = 16 = 4^2; 4^3 = 64, 4^6 = 4096 = 64^2$, etc. We can gain an understanding of fractional exponents by considering the reverse of this process: halving is the reverse of doubling, and taking the square root is the reverse of squaring. It seems reasonable therefore to define $16^{\frac{1}{2}}$ as 4, $4096^{\frac{1}{2}}$ as 64, or more generally $a^{\frac{1}{2}} = \sqrt{a}$. Similarly, if $a^3 = b$, then $b^{\frac{1}{3}} = a$; if $a^4 = b$, then $b^{\frac{1}{4}} = a$. In this way the logic can be extended until *any* number can be used as an exponent: for example, $2^{0.21} = 2^{21/100} = (2^{21})^{1/100} = 2\,097\,152^{1/100} = 1.1567$ approximately.

If a is a positive number (not necessarily an integer), then a^i is meaningful for *any* value of i, as we have seen. But if a is negative, a^i has a simple meaning only if i is a positive or negative *integer*, and in this case the meaning is just the same as for positive a. For example:

$$(-2)^1 = -2$$
$$(-2)^2 = (-2)(-2) = +4$$
$$(-2)^3 = (-2)(-2)(-2) = -8$$
$$(-2)^{-1} = 1/(-2) = -\tfrac{1}{2}, \text{ etc.}$$

[I shall not consider fractional powers of negative numbers, such as $(-2)^{\frac{1}{2}}$, because, although there are circumstances in which it is useful to define such powers, these circumstances occur very infrequently in biochemistry: for the purposes of this book, therefore, we can insist that i must be integral for a^i to have a meaning if a is negative.]

In the above set of examples, notice that $(-2)^2 = +4 = 2^2$, and in general, for any a, $(-a)^2 = a^2$. This raises the possibility of ambiguity in the definition of fractional powers. If we define $4^{\frac{1}{2}}$ as 'the number that when multiplied by itself gives 4', the definition is ambiguous because it is satisfied not only by 2 but also by -2. In most circumstances it is not at all convenient for a mathematical expression to have more than one meaning, and with fractional powers the ambiguity is removed by convention: $b^{\frac{1}{2}}$ is defined as the *positive* number a that satisfies the relationship $a^2 = b$. If we want to refer to the negative solution to the equation $a^2 = b$ we must write $-b^{\frac{1}{2}}$. Occasionally we want to retain the ambiguity (as in the formula for the roots of a quadratic equation: see Chapter 5), and in this case we use the symbol $\pm$ ('plus or minus') to indicate that either sign may be used.

2.3   Addition and subtraction of exponents

We have seen that a^i means the product of i a's, and similarly a^j means the product of j a's. If these two products are multiplied together, the result must be the same as if we had multiplied $(i+j)$ a's together at the outset, i.e. a^{i+j}. In general:

$$a^{i+j} = a^i a^j$$

Although this result is obviously correct for integer values of i and j only, it actually applies for any values of i and j, including both fractional and negative ones. If j is negative – suppose $j = -k$, where k is positive – then the expression above gives

$$a^{i+j} = a^{i-k} = a^i a^j = a^i a^{-k} = a^i/a^k$$

Thus, just as *addition* of exponents corresponds to *multiplication* of powers, so *subtraction* of exponents corresponds to *division* of powers.

It is important to note that in all of the expressions in this section I have used the same base a: there are no correspondingly simple expressions for powers of *different* numbers. For example, if $a \neq b$, $a^i b^j$ does not have a simple expression in terms of added or subtracted exponents (unless of course b is a simple power of a, e.g. if $b = a^2$ then $a^i b^j = a^{i+2j}$).

2.4 Logarithms

Addition and subtraction are much quicker and easier operations to carry out than multiplication and division. The observation that addition and subtraction of exponents corresponds to multiplication and division of powers suggests a way in which one might use the operations of addition and subtraction to obtain the results of multiplication and division. For example, if one had a list of powers of 2, one might multiply 4 by 8 by noting that $4 = 2^2$ and $8 = 2^3$, so $4 \times 8 = 2^{2+3} = 2^5 = 32$. This is of course a trivial example as it stands, because 4 can be multiplied by 8 in the head more rapidly than one can consult a lists of powers of 2. Moreover, if the list were confined to *integral* powers it would be unlikely to include the numbers we wished to multiply together. Suppose, however, that the list were extensive and included non-integral exponents: then we could use the same approach to solve non-trivial problems. Consider, for example, the product 2.38×9.23. An extensive table of powers of 2 would reveal that $2.38 \simeq 2^{1.25}$ and $9.23 \simeq 2^{3.21}$; therefore, $2.38 \times 9.23 \simeq 2^{1.25 + 3.21} = 2^{4.46} \simeq 21.97$. In this case the direct multiplication is not trivial and would take much more time than the corresponding calculation in terms of powers of 2 – provided, of course, that suitable tables were available. In practice it would be most convenient to have *two* sets of tables, one showing what exponent i is needed for 2^i to have a selected value, the second showing the value of 2^i for a selected value of i. These are known as tables of *logarithms* and *antilogarithms*, respectively. Just as 2.38 can be described as the result of raising 2 to the power 1.25, so we can describe 1.25 as the logarithm of 2.38 to the base 2. These relationships are symbolized as follows:

$$1.25 = \log_2 2.38$$
$$2.38 = \text{antilog}_2 1.25 = 2^{1.25}$$

Notice that 'antilogarithm' is simply another word for power: it is used to emphasize that finding an antilogarithm or a power is simply the reverse of finding a logarithm.

2.5 Common logarithms

It is not necessary to use 2 as the base for a set of logarithms: any positive number would in principle be usable, and it might seem that all positive numbers would be equally convenient to use. In fact, however, the two bases 10 and e (see p. 31, natural logarithms) have considerable advantages over all other choices and are by far the most commonly used, although 2 is sometimes used in studies of bacterial growth, for reasons that I shall discuss later in this chapter. The practical advantage of 10 over, say, 2 is apparent from the following comparison of $\log_{10}$ and $\log_2$ values for several numbers:

$$\log_2 2.741 = 1.454; \qquad \log_{10} 2.741 = 0.438$$
$$\log_2 27.41 = 4.776; \qquad \log_{10} 27.41 = 1.438$$
$$\log_2 274.1 = 8.098; \qquad \log_{10} 274.1 = 2.438$$
$$\log_2 2741 = 11.420; \qquad \log_{10} 2741 = 3.438$$

Notice that each multiplication of a by 10 increases $\log_2 a$ by $\log_2 10 = 3.322$, a rather unmemorable number, but $\log_{10} a$ by $\log_{10} 10 = 1$ exactly. This means that if we use logarithms to the base 10, or *common logarithms* (also known sometimes a Briggsian logarithms, after Henry Briggs, the English mathematician who first published tables of them in 1617), we need only a limited set of tables of $\log_{10} a$ for a values from 1 to 10, because all other common logarithms can be obtained from these by a trivial calculation. For example, if we required $\log_{10} 274.1$, we would look up $\log_{10} 2.741 = 0.438$ in a set of tables, note that 274.1 is $100 = 10^2$ times 2.741 and add 2 to the logarithm of 2.741, so $\log_{10} 274.1 = 2.438$.

Because of the separate ways of finding the integral and non-integral parts of a logarithm, they are given different names. The integral part (before the decimal point) is called the *characteristic*; the non-integral part (after the decimal point) is called the *mantissa*. Tables of common logarithms provide only the mantissa, as the characteristic is easy to determine by inspection. Conversely, tables of antilogarithms provide values only for the antilogarithm of the mantissa, the nearest power of ten being found by inspection of the characteristic.

With negative logarithms, which are the logarithms of numbers between 0 and 1 (*not* the logarithms of negative numbers, which do not have logarithms: see below), it is common but not universal practice to retain a positive mantissa. For example, $\log_{10} 0.2741 = -0.562$, but if it is written in this way it is not immediately obvious that it is exactly 1 less than $\log_{10} 2.741 = 0.438$. Instead, therefore, it can be written as $-1 + 0.438$ or, more concisely, as $\overline{1}.438$ (spoken aloud as 'bar 1 point 438'), where the minus sign is written *above* the characteristic to show that it applies only to the characteristic and not to the mantissa. Similarly, $\log_{10} 0.002\,741 = -3 + 0.438 = \bar{3}.438$, etc. The need for this notation has decreased with the widespread availability of computers and electronic calculators, which have removed much of the drudgery from arithmetic, and have in particular made multiplication and division into trivial exercises. Consequently one now has little occasion to write quantities such as $\bar{3}.438$, and one should especially avoid them in preparing data for plotting graphs, where they cause many mistakes. Nonetheless the student should be familiar with the meaning of the notation as it is still encountered occasionally.

We have seen that $a^0 = 1$, regardless of the value of a. It follows that $\log_a 1 = 0$, regardless of a. The logic of this is evident from the fact that 0 is the only number that can be *added* to another number without affecting its value, whereas 1 is the only number that can be *multiplied* by another number without affecting its value.

The following equations summarize this section and the preceding one:

$$\log(ab) = \log a + \log b$$
$$ab = \text{antilog}(\log a + \log b)$$

I have omitted the base of the logarithms from the symbols in these equations, because they apply for all bases – so long as the same base is used consistently: it would be improper, for example, to mix logarithms to bases 10 and e in the same equation. The usual practice in biochemistry is to use the symbol log to mean $\log_{10}$, and ln to mean $\log_e$. Be warned, however, that this practice is not universal, and that pure mathematicians and others sometimes use log to mean $\log_e$.

2.6 Negative numbers have no logarithms

Each division of a number by 10 causes a decrease in its logarithm to the base 10 of exactly 1. No amount of dividing by 10 can turn a positive number into a negative number, however, and so the original number remains positive throughout, even though its logarithm can cross zero and become negative. It is clear, therefore, that negative logarithms refer to small positive numbers, not to negative numbers. Negative numbers cannot be expressed as powers of positive numbers, and consequently have no logarithms. This does not mean, however, that it is impossible to evaluate products that involve negative numbers. For example, -1.274×3.859 can readily be evaluated by use of logarithms by first evaluating 1.274×3.859 and then changing the sign of the result.

2.7 Natural logarithms

The second widely used base of logarithms is at first sight a bizarre choice, the number e. This number is *irrational*, which means that it cannot be expressed as an exact ratio of two integers, and consequently cannot be written exactly as a decimal number. It can, however, be expressed to any required accuracy as a decimal, and for most purposes it is more than sufficiently accurate to express its value as 2.718 28. One reason for choosing e as a base for logarithms is that it is very easy to calculate its non-integral powers, by means of the following series:

$$e^x = 1 + \frac{x}{1!} + \frac{x^2}{2!} + \frac{x^3}{3!} + \frac{x^4}{4!} + \cdots$$

in which the exclamation point indicates a *factorial*, i.e. $n! = n(n-1)$ $(n-2) \ldots 3 \times 2 \times 1$; for example, $6! = 6 \times 5 \times 4 \times 3 \times 2 \times 1 = 720$. Although the series for e^x is strictly an *infinite* series, with an infinite number of terms, it is never necessary to evaluate more than a finite number of them, because the magnitude of the terms always dwindles into insignificance, after the first few terms, unless x is very large. If x is less than 1 this decay into insignificance is very rapid and for small values of x it is sometimes useful to consider only the first two terms in the series:

$$e^x \simeq 1 + x \text{ (accurate to within } \pm 5\% \text{ if } -0.28 < x < 0.35)$$

Although the infinite series provides a convenient way of calculating e^x, i.e. antilog$_e x$, rather than log$_e x$, a table of e^x values can be read inversely as a table of log$_e x$ values. It is much easier, therefore, to obtain logarithms and antilogarithms with e as base than with 10 and other bases. Nonetheless, once a set of tables with base e is available it is in principle simple to create a corresponding set with base 10, because for any number x there is a simple relationship between the logarithms to the different bases, as follows:

$$\log_e x = \log_e 10 \log_{10} x = 2.303 \log_{10} x$$

As we have seen, however, 10 is much the most convenient base for purposes of calculation: why then are logarithms to base e of more than just curiosity interest?

The continued importance of e and its functions such as $\log_e x$ and e^x does not, in fact, derive from their almost nonexistent use as aids to arithmetic but from the fact that they occur naturally in the solutions to many kinds of problems in applied mathematics. We shall encounter various examples of this in this book and at this point it will suffice to indicate two briefly:

(1) In a system at thermal equilibrium at an absolute temperature T, the numbers n_1 and n_2 of molecules in two states with energies E_1 and E_2, respectively, are related according to the equation

$$n_1/n_2 = e^{(E_2 - E_1)/kT}$$

in which k is a constant known as the Boltzmann constant $(= 1.38 \times 10^{-23} \, \text{J K}^{-1} = R/N$, where $R = 8.31 \, \text{J K}^{-1} \, \text{mol}^{-1}$ is the gas constant and $N = 6.02 \times 10^{23} \, \text{mol}^{-1}$ is Avogadro's number).

(2) In a first-order reaction with rate constant k the extent of reaction after time t is proportional to $1 - e^{-kt}$.

Although I have deliberately chosen rather physical examples here to illustrate the ubiquity of e, the fundamental reasons for the ubiquity are mathematical: they derive from the natural occurrence of $\log_e x$ and e^x in purely mathematical contexts, as we shall see when examining calculus (Chapters 3–4).

Powers of e occur so often that it is typographically inconvenient to write them as e^x, especially if 'x' is a complicated expression. Accordingly the special operator exp is commonly used instead, i.e.

$$\exp(x) \equiv e^x$$

the expression being spoken aloud as 'exponential x': it is unnecessary to specify the base because e is universally understood in this context.

2.8   Logarithms to base 2

We have seen earlier in this chapter that logarithms to base 2 provide a convenient and easily understood introduction to the idea of logarithms as an aid to arithmetic. Although for most practical purposes 10 is the easiest base to work with and e is usually appropriate in theoretical discussions, 2 has an important use as a base in bacteriology.

When bacteria grow in ideal conditions in an open system provided with ample supplies of all necessary nutrients, the rate of growth is determined only by the metabolic capability of each individual bacterium to grow and divide. In a homogeneous bacterial culture there are no qualitative differences between individuals and quantitative differences are slight. Consequently it is meaningful to conceive of the *generation time*, i.e. the period between one cell division and the next, as fairly constant from one cell to another. If the 'initial' population size (i.e. the number of cells at an arbitrarily defined time zero) is n_0 and the generation time is τ, then at

$$
\begin{aligned}
t = 0, & \quad n = n_0 \\
t = \tau, & \quad n = 2n_0 \\
t = 2\tau, & \quad n = 4n_0 \\
t = 3\tau, & \quad n = 8n_0
\end{aligned}
$$

and, in general, at

$$ t = z\tau, \; n = 2^z n_0 $$

Taking logarithms (to any base), we have

$$ \log n = z \log 2 + \log n_0 $$

or

$$ t/\tau = \frac{\log (n/n_0)}{\log 2} = \frac{\log n - \log n_0}{\log 2} $$

Although this equation is independent of the base of the logarithms it assumes a particularly simple form when the base is 2, because $\log_2 2 = 1$, so

$$ t/\tau = \log_2 (n/n_0) = \log_2 n - \log_2 n_0 $$

The phase of growth in which the mean generation time is a constant is known properly as the *exponential growth phase*. Paradoxically, the term *logarithmic growth phase* is synonymous, even though it sounds as if it means the opposite. The reason for this peculiar terminology is that during exponential growth the population size is not a linear function of time but can be plotted as a straight line if its *logarithm* is plotted against time (Fig. 2.1)

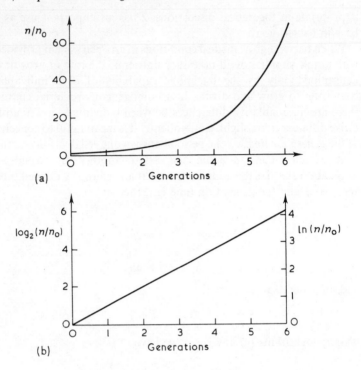

Fig. 2.1.   Logarithms to base 2 provide a convenient way of relating population size to number of generations, for populations that double in size for each generation.

2.9   Exponential decay

Although exponential *growth* is mainly observed with bacteria, exponential *decay* is of great importance to the biochemist in at least two other contexts: radioactive decay and chemical reactions that display first-order kinetics. Mathematically, all of these processes are very similar, differing only in the sign of the coefficient of time in the exponent, and the mean generation time in a growth process is parallelled by the *half life* of a radioisotope or reactive chemical and the *half time* of a chemical reaction. In principle, logarithms to base 2 might be used in these contexts but in practice they are not. The reason for this is partly historical: methods for analysing decay processes were already well established before 1949, when Monod drew attention to the usefulness of 2 as a base; but more important is the fact that the mean generation time of a bacterial population has a tangible physical meaning whereas the half life of a radioisotope is only a mathematical abstraction, albeit a convenient one. Thus there is a real interest in knowing the mean number of generations in a period of bacterial growth, whereas the number of times that a sample has lost half of its radioactivity is no more fundamental than the number of times its activity has decreased by a factor of e or by a factor of 10.

2.10 Logarithms as a method of scaling

In the chemistry of processes in aqueous solution the concentrations of hydrogen ions that need to be considered extend over more than 14 orders of magnitude, from less than 10^{-14} mol l^{-1} in strong alkali to more than 1 mol l^{-1} in strong acid. In the chemistry of living systems this range is considerably compressed but it is still large enough to make it arithmetically inconvenient to work exclusively in concentrations. In addition to the numerical inconvenience of concentrations there is a philosophical reason for avoiding them. Biochemistry – and indeed much of chemistry – tends to be more interested in changes in state than in states, and a particular change in concentration can have very different effects in different contexts. For example, in the mammalian stomach at a hydrogen-ion concentration of about 10^{-3} mol l^{-1} an increase of 10^{-5} mol l^{-1} would be virtually imperceptible and would be expected to have only slight consequences; on the other hand the same increase in a cell at a hydrogen-ion concentration of 10^{-7} mol l^{-1} might well be devastating. (This is not a wholly abstract example: the increase in acidity brought about by rupturing lysosomes has major effects on the activities of numerous enzymes within the cell but would pass unnoticed in the stomach). The point is that in the first case the change is 1.01-fold whereas in the second it is 100-fold: in most circumstances it is equal *relative* changes that bring about similar effects. It would be an oversimplification to say that a doubling (for example) of hydrogen-ion concentration has the 'same' effect when applied to starting values of 10^{-3} and 10^{-7} mol l^{-1}, but such a statement makes sense conceptually and is not without meaning. On the other hand it is difficult to think of circumstances in which it would make sense to regard increases of 10^{-5} mol l^{-1} as the 'same', whether from 10^{-7} or from 10^{-3} mol l^{-1}.

The purpose of this lengthy digression has been to show that there are reasons other than numerical convenience for introducing logarithms into consideration of hydrogen-ion concentration. Not only does the conversion of $[H^+]$ into $\log_{10}[H^+]$ avoid the need for inconveniently small numbers; it also provides a scale in which particular linear changes (in $\log_{10}[H^+]$) have approximately constant meanings. Unfortunately the scale that is in common use is not $\log_{10}[H^+]$ but $-\log_{10}[H^+]$, generally written as pH. For the ludicrously trivial 'advantage' of having a range of positive numbers around $+7$ rather than negative numbers around -7, chemists have paid the excessive price of discarding a scale of obvious meaning in favour of one that has generated endless confusion. The pH scale is probably here to stay, however, and so one must make the best of an error of judgement and try to remember that pH is nothing more than a tiresome way of writing $-\log_{10}[H^+]$. Similarly pK_a means $-\log_{10}K_a$, where K_a is an acid dissociation constant. Other usages, such as pK_m (where K_m is the Michaelis constant) are less common: I trust that they will remain so and that the time will never come when it is thought helpful to overcome the 'inconvenience' of having a minus sign in the definition of the standard Gibbs energy

$$\Delta G^\circ = -RT \ln K$$

by writing it as $\Delta G^\circ = 2.3RTpK$. We are fortunate that at least some extensions of the pH idea, such as the rH scale for application to redox potentials, have become much less prevalent than they once were.

2.11 Products of equilibrium constants

In a set of equilibria for a sequence of reactions, such as the following sequence of unimolecular steps:

$$A \underset{}{\overset{K_{AB}}{\rightleftarrows}} B \underset{}{\overset{K_{BC}}{\rightleftarrows}} C \underset{}{\overset{K_{CD}}{\rightleftarrows}} D \underset{}{\overset{K_{DE}}{\rightleftarrows}} E \underset{}{\overset{K_{EF}}{\rightleftarrows}} F$$

one can obtain the equilibrium constant for the complete process by multiplying all of the individual equilibrium constants together:

$$K_{AF} = K_{AB}K_{BC}K_{CD}K_{DE}K_{EF}$$

Metabolic pathways are composed of such sequences of steps

(although many are more complex than unimolecular steps) and it is frequently necessary to carry out such multiplications and corresponding divisions in order to assess the thermodynamic feasibility of particular pathways or importance of intermediates, etc. But it is much easier to think in terms of addition and subtraction than in terms of multiplication and division and for this reason it is often useful to express equilibrium constants on a logarithmic scale, so that

$$\log K_{AF} = \log K_{AB} + \log K_{BC} + \log K_{CD} + \log K_{DE} + \log K_{EF}$$

Now it happens, for reasons that are primarily chemical rather than mathematical, that there is a direct relationship between the equilibrium constant of a reaction and the enthalpy and entropy changes that occur in it. Because of this relationship, it is common practice to use not the simple logarithm to base 10 – which would be best if combining equilibrium constants were the only concern – but instead the natural logarithm multiplied by a temperature-dependent coefficient, i.e. the standard Gibbs energy ΔG°, defined as follows:

$$\Delta G^\circ = -RT \ln K = -2.3RT \log_{10} K$$

where $R = 8.31$ J mol^{-1} is the gas constant and T is the temperature in kelvins. Nonetheless although ΔG° has a fundamental thermodynamic meaning the factor $-2.3RT$ is irrelevant to many of the biochemical contexts in which a logarithmic scale is useful. Living systems hardly ever exploit temperature effects – indeed they usually try to avoid them altogether – and so in metabolism $-2.3RT$ is just a constant, with a value of about -5700 J mol^{-1} (at 25°C). Thus each -5.7 kJ mol^{-1} in a ΔG° value corresponds to a factor of 10 in the corresponding equilibrium constant.

2.12 Logarithms of dimensioned quantities?

It is obvious from consideration of the simplest uses of exponents that they must be pure numbers, i.e. they must not have any dimensions: it is meaningless to multiply 10 by itself 3 cm times, for example. As a logarithm is the exponent that the base must be raised to to give a particular number it follows that a logarithm can have no dimensions. Moreover, a pure number raised to a pure-number power must also be a pure number, and so only pure numbers have logarithms.

Despite this, there is apparently a glaring exception, one that has already been considered in this chapter: if we define pH as $-\log_{10}[H^+]$, and we measure $[H^+]$ in $mol\,l^{-1}$ then we apparently must take the logarithm of a concentration. The way out of this difficulty is to define a *standard state*, in this instance a reference concentration $[H^+]^0$, and to define pH as $-\log_{10}([H^+]/[H^+]^0)$. In this way pH becomes the logarithm of a dimensionless ratio, as it should be. In practice we take the standard state as $[H^+]^0 = 1\,mol\,l^{-1}$ and thereby make the ordinary definition of pH numerically correct *provided that* $[H^+]$ *is measured in* $mol\,l^{-1}$.

The fuller definition of pH may seem pedantic and trivial, but it is important for understanding that although pH *differences* have a fundamental meaning, pH itself is measured from an arbitrary datum and therefore has no absolute meaning. Thus although it makes good sense to say that a pH change of $+0.2$ is twice as large as a change of $+0.1$, there is no useful sense in which pH 6 can be regarded as twice as alkaline (or even as twice as 'large', where the definition of 'large' is kept suitably vague) as pH 3. Suppose, for example, that we chose to redefine the standard state as $[H^+]^{0'} = 10^{-7}\,mol\,l^{-1}$ (a datum that would make better biochemical sense than the chemists' $1\,mol\,l^{-1}$): in this case pH 3 would become $pH' = -4$ and pH 6 would become $pH' = -1$, a quarter as 'large' instead of twice as 'large'!; but a *change* in pH of $+0.2$ would also be a change in pH' of $+0.2$ and would still be twice as large as $+0.1$, because the standard state does not enter into measurements of change.

Not many biochemists would suppose that pH 6 could meaningfully be regarded as twice as large as pH 3, but equally ridiculous comparisons are commonplace in discussions of ΔG° values for metabolic processes. Standard states do not enter into the definitions of dimensionless equilibrium constants (as in the reversible unimolecular reactions considered in the preceding section), but many metabolic and other reactions involve changes in the number of molecules and have equilibrium constants with dimensions that must be removed by the use of standard states. As in the definition of pH, the commonest convention is to define dimensionless equilibrium constants by relating concentrations to standard concentrations of $1\,mol\,l^{-1}$. As a consequence of this introduction of standard states it is nearly always meaningless to measure the 'efficiencies' of metabolic processes by dividing ΔG° values into one another.

2.13 Redox potentials*

A *redox couple* is a pair of substances such that one, the oxidized form, can be converted into the other, the reduced form, by accepting one or more electrons. For example, the $Fe^{3+}|Fe^{2+}$ couple can be written as

$$Fe^{3+} + \varepsilon^- \rightarrow Fe^{2+}$$

Now free electrons cannot exist in aqueous solution and so this reaction, a so-called *half-reaction*, cannot occur unless there is another half-reaction occurring simultaneously in the reverse direction. Nonetheless, certain couples occur in so many different metabolic reactions that it is useful to be able to express their oxidizing ability without explicit reference to another half-reaction. This may be done by means of the *redox potential E*, which is defined by the *Nernst equation*:

$$E = E^0 + \frac{RT}{n\mathscr{F}} \ln \frac{[\text{ox}]}{[\text{red}]}$$

in which E^0 is a constant for the particular couple known as the *standard redox potential*, $R = 8.314 \, \text{J mol}^{-1} \text{K}^{-1}$ is the gas constant, T is the temperature in kelvins, n is the number of electrons transferred, $\mathscr{F} = 96\,495 \, \text{coulomb mol}^{-1}$ is a constant that allows E and E^0 to be measured in volts, and [ox] and [red] are, respectively, the concentrations of the oxidized and reduced forms of the couple.

For the $Fe^{3+}|Fe^{2+}$ couple, $E^0 = 0.77$ volt and $n = 1$, so

$$E = 0.77 + \frac{RT}{\mathscr{F}} \ln \frac{[Fe^{3+}]}{[Fe^{2+}]} = 0.77 + 0.060 \log \frac{[Fe^{3+}]}{[Fe^{2+}]}$$

where the coefficient 0.060 V is the value of $2.303 \, RT/\mathscr{F}$ at 303 K (30° C). This tells us that the actual potential E is determined by two separate quantities, not only by the standard potential E^0 but also by the ratio of concentrations of the two components: not surprisingly, a couple becomes more powerfully oxidizing as the proportion in the oxidized state increases. Rather confusingly, the Nernst equation is

---

* Strictly the term is *oxidation-reduction potential*, but this is rather cumbersome and I shall use the shorter and more colloquial form in this discussion.

sometimes written with a negative sign and the ratio inverted:

$$E = E^0 - \frac{RT}{n\mathscr{F}} \ln\frac{[\text{red}]}{[\text{ox}]}$$

It should be evident from the discussion of the properties of logarithms earlier in this chapter that this form is exactly equivalent. Confusion can be avoided by remembering that whichever way the equation is written increasing the proportion of *reduced* form *reduces* the value of the potential.

The meaning of E^0 can be understood by putting $[\text{red}] = [\text{ox}]$ $= 1 \text{ mol} \, l^{-1}$. (Actually the particular concentration is not important for the mathematical argument, which would be unchanged if $[\text{red}]$ $= [\text{ox}]$ at some concentration other than $1 \text{ mol} \, l^{-1}$. However, this concentration is specified for defining the standard states and in practice it is desirable that it should be specified because the Nernst equation may not be obeyed exactly.) Then the logarithmic term becomes zero and so

$$E = E^0$$

Thus E^0 is the value of E when the components of the couple are in their standard states. The absolute value of E^0 is arbitrary because it is measured from an arbitrary zero. (The value of E^0 for one particular couple, the *standard hydrogen electrode* $H^+ | \frac{1}{2} H_2$ under specified conditions is arbitrarily *defined* as zero, and all other E^0 values are defined in relation to this standard). Consequently E^0 values (and E values, for that matter) for different couples cannot be compared as ratios but must only be compared as differences (compare the discussion of pH and dimensions in the preceding section).

The $Fe^{3+} | Fe^{2+}$ couple is particularly simple as an introduction to redox potentials because no protons are involved in the half-reaction. Consequently its potential is independent of the hydrogen-ion concentration. In fact E^0 is defined at pH 0, although for the $Fe^{3+} | Fe^{2+}$ couple it would have exactly the same value of 0.77 V if it were defined at pH 7 or some other pH. This is not the case for most couples of biochemical interest, because most involve a transfer of protons as well as electrons. If protons are transferred, but neither form of the couple ionizes in the pH range of interest, they may be allowed for by including them in the Nernst equation as reactants. For example, the acetaldehyde|ethanol couple involves two protons:

$$CH_3CHO + 2H^+ + 2\varepsilon^- \rightarrow C_2H_5OH$$

and has $E^0 = 0.258$ V. So the potential is given by

$$E = 0.258 + \frac{RT}{2\mathscr{F}} \ln \frac{[CH_3CHO][H^+]^2}{[C_2H_5OH]}$$

$$= 0.258 + \frac{RT}{2\mathscr{F}} \ln \frac{[CH_3CHO]}{[C_2H_5OH]} + \frac{RT}{\mathscr{F}} \ln [H^+] \text{ (volt)}$$

Remember that dividing a logarithm by 2 is equivalent to taking the square root of the quantity whose logarithm is expressed. So if we remove the 2 from the coefficient we can replace $[H^+]^2$ with $[H^+]$. Furthermore, $RT/\mathscr{F} = 0.060$ V, and so

$$E = 0.258 + 0.030 \log \frac{[CH_3CHO]}{[C_2H_5OH]} - 0.060 \, pH \text{ (volt)}$$

Thus at any $[CH_3CHO]/[C_2H_5OH]$ ratio the value of E is 0.42 V less at pH 7 than at pH 0.

Of the couples of biochemical importance, some, for example cytochrome c(Fe^{3+})|cytochrome c(Fe^{2+}), resemble Fe^{3+}|Fe^{2+} in having redox potentials that are independent of pH; a few, such as acetaldehyde|ethanol, show a decrease of 0.06 V for each unit increase in pH; but many show intermediate behaviour, because one or both components ionize in the pH-range of interest (as I shall discuss in the next section). It follows that E^0 values if applied naïvely can be a very misleading guide to redox behaviour at neutral pH values. For example, cytochrome c(Fe^{3+})|cytochrome c(Fe^{2+}) and acetaldehyde|ethanol have nearly equal E^0 values, 0.250 V and 0.258 V, respectively, but under metabolic conditions the former is a much more powerfully oxidizing couple than the latter, because of the large decrease in potential shown by the latter on raising the pH. To avoid this confusion biochemists commonly redefine the standard as the observed potential at a specified pH (usually pH 7) when the two components are at equal concentrations. This potential, often called the *mid-point potential*, is given the symbol $E^{0\prime}$, and the Nernst equation becomes

$$E = E^{0\prime} + \frac{RT}{n\mathscr{F}} \ln \frac{[ox]}{[red]}$$

Proton concentrations are *not* included in this equation and so it is valid only at the pH at which $E^{0\prime}$ is defined. At pH 7 we have $E^{0\prime} = -0.163$ V for acetaldehyde|ethanol, but still $E^{0\prime} = E^0 = +0.250$ V for cytochrome $c(Fe^{3+})$|cytochrome $c(Fe^{2+})$.

If two redox couples are allowed to interact, whether by being mixed together in the presence of a suitable catalyst or by connecting them electrically, they can react stoichiometrically until their potentials are identical. For example, in the reaction catalysed by alcohol dehydrogenase the acetaldehyde|ethanol couple reacts with the NAD^+|NADH couple:

$$NAD^+ + H^+ + 2\varepsilon^- \rightarrow NADH \qquad E^{0\prime} = -0.320 \text{ V (pH 7, 30°C)}$$

When the two E values are equal, i.e. at equilibrium, we can write

$$-0.163 + 0.03 \log \frac{[CH_3CHO]}{[C_2H_5OH]} = -0.320 + 0.03 \log \frac{[NAD^+]}{[NADH]}$$

which may be rearranged to give

$$K = \frac{[NAD^+][C_2H_5OH]}{[NADH][CH_3CHO]} = 1.71 \times 10^5$$

which is the equilibrium constant for the reaction at pH 7 and 30°C.

Like the standard Gibbs energy of reaction $\Delta G^{0\prime}$, the difference between two standard redox potentials is a logarithmic expression of an equilibrium constant, and in general we can express this difference $\Delta E^{0\prime}$ as follows:

$$\Delta E^{0\prime} = \frac{-RT}{n\mathscr{F}} \ln K = \Delta G^{0\prime}/n\mathscr{F}$$

Thus standard redox potential differences and standard Gibbs energy of reaction are measurements of the same thing but in different units, volts for $\Delta E^{0\prime}$ and $kJ \, mol^{-1}$ for $\Delta G^{0\prime}$. The reason why redox potentials are commonly given in electrical units is that electrochemical cells often provide the most convenient and accurate way of determining equilibrium constants, especially for reactions in which the equilibrium position is very far in one direction.

2.14 Dependence of redox potentials on pH

For many redox couples one or sometimes both of the components ionize between pH 0 and pH 7 and consequently E varies with pH in a more interesting way than in either of the simple examples of pH behaviour discussed in the preceding section. Consider, for example, the couple acetic acid|acetaldehyde:

$$CH_3CO_2H + 2H^+ + 2\varepsilon^- \rightarrow CH_3CHO + H_2O$$

for which $E^0 = -0.11$ V and the pK_a ($= -\log K_a$) of the oxidized form acetic acid is 4.73. Omitting water (i.e. following the convention that the standard state of water is the state that exists in the solution of interest), the Nernst equation can be written as follows:

$$E = E^0 + \frac{RT}{2\mathscr{F}} \ln \frac{[CH_3CO_2H]}{[CH_3CHO]} + \frac{RT}{\mathscr{F}} \ln [H^+]$$

i.e. in a form analogous to that used for the acetaldehyde|ethanol couple considered in the previous section. However, for the acetic acid|acetaldehyde couple this expression is inconvenient to use because it contains the concentration of *protonated* oxidized form, i.e. $[CH_3CO_2H]$, even though experimentally it would be difficult to maintain this concentration constant as the pH changed. In reality we would usually keep the *total* concentration of oxidized form, $[CH_3CO_2(H)] = [CH_3CO_2H] + [CH_3CO_2^-]$, constant; so it would be much more helpful to recast the Nernst equation in terms of this total concentration, which is given by

$$[CH_3CO_2(H)] = [CH_3CO_2H](1 + K_a/[H^+])$$

and so

$$E = E^0 + \frac{RT}{2\mathscr{F}} \ln \frac{[CH_3CO_2(H)]}{[CH_3CHO](1 + K_a/[H^+])} + \frac{RT}{\mathscr{F}} \ln [H^+]$$

$$= E^0 + \frac{RT}{2\mathscr{F}} \ln \frac{[CH_3CO_2(H)]}{[CH_3CHO]} + \frac{RT}{\mathscr{F}} \ln [H^+]$$

$$- \frac{RT}{2\mathscr{F}} \ln (1 + K_a/[H^+])$$

Then the mid-point potential at any pH is given by

$$E^{0\prime} = E^0 + \frac{RT}{\mathscr{F}} \ln [H^+] - \frac{RT}{2\mathscr{F}} \ln (1 + K_a/[H^+])$$

$$= -0.11 - 0.06\,pH - 0.03\log(1 + K_a/[H^+]) \quad (volt)$$

The general form of this expression may be found by considering its limits as $[H^+]$ is very large or very small. At low pH, $K_a/[H^+]$ is negligible compared with 1, and so the third term approximates to $0.03 \log 1$, i.e. zero, and

$$E^{0\prime} \simeq -0.11 - 0.06\,\mathrm{pH} \quad (\text{volt}) \quad ([H^+] \gg K_a)$$

At high pH, 1 is negligible compared with $K_a/[H^+]$, and so

$$E^{0\prime} \simeq -0.11 - 0.06\,\mathrm{pH} - 0.03 \log (K_a/[H^+])$$
$$= -0.11 - 0.06\,\mathrm{pH} + 0.03\,\mathrm{p}K_a - 0.03\,\mathrm{pH}$$
$$= +0.03 - 0.09\,\mathrm{pH} \quad (\text{volt}) \qquad ([H^+] \ll K_a)$$

Finally, one can form an impression of the behaviour between the two limiting regions by putting $[H^+] = K_a$:

$$E^{0\prime} = -0.11 - 0.06\,\mathrm{p}K_a - 0.03 \log 2 = -0.40\,\mathrm{V} \quad ([H^+] = K_a)$$

From these last three expressions it is clear that the dependence of the mid-point potential on pH is as shown in Fig. 2.2: at low pH, $E^{0\prime}$ decreases linearly by 0.06 V for each unit increase in pH; at high pH it decreases linearly by 0.09 V for each unit increase in pH; the two linear

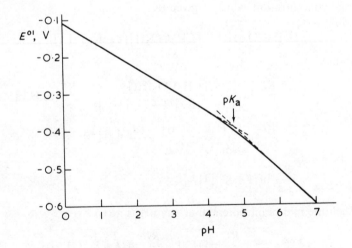

Fig. 2.2.   Dependence of mid-point potential on pH for the acetic acid|acetaldehyde couple.

regions are joined by a curve, but the extrapolated straight lines intersect at $pH = pK_a$.

I have discussed redox potentials and their dependence on pH in rather more detail than one would usually expect in an elementary mathematics book because it is a topic that causes a great deal of misunderstanding and muddle, not only among students but also, regrettably, among the authors of biochemistry textbooks. Most of this misunderstanding can be avoided by realizing the following: (i) the Nernst equation at pH 0 is normally expressed in terms of a single oxidizing form and a single reducing form; (ii) in many couples of

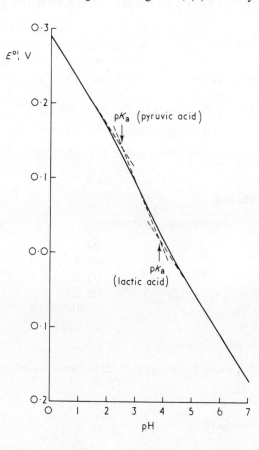

Fig. 2.3.   Dependence of mid-point potential on pH for the pyruvic acid|lactic acid couple.

interest ionization of one or both forms occurs between pH 0 and pH 7; (iii) in ordinary experiments one controls the *total* concentrations of oxidized and reduced forms not the concentrations of any particular ionic states; (iv) one therefore needs to re-express the Nernst equation in terms of total concentrations; (v) this can be done by means of the usual equations for ionic equilibria. By following these rules it is quite straightforward to deal with couples in which more than a single ionization must be considered. For example, the couple pyruvic acid|lactic acid,

$$CH_3COCO_2H + 2H^+ + 2\varepsilon^- \rightarrow CH_3CHOHCO_2H$$

has $E^0 = 0.29\,V$, $pK_{a(pyr)} = 2.50$, $pK_{a(lact)} = 3.86$ and thus requires ionizations of both oxidized and reduced forms to be taken into account. Nonetheless, a derivation along the lines of that shown above for acetic acid|acetaldehyde is not difficult and leads to

$$E^{0\prime} = 0.29 - 0.06\,pH - 0.03\,\log(1 + K_{a(pyr)}/[H^+])$$
$$+ 0.03\,\log(1 + K_{a(lact)}/[H^+]) \quad (volt)$$

and the dependence on pH shown in Fig. 2.3.

2.15 Problems

(2.1) Evaluate the following expressions:

(a) 3^3 (b) 4^{-2}
(c) $27^{\frac{1}{3}}$ (d) $4^{-\frac{1}{2}}$
(e) 16^0 (f) 10^{-3}
(g) $3.29^6 \times 3.29^{-4} \times 3.29^{-3} \times 3.29$ (h) $\overline{2}.37$
(i) $4!$ (j) antilog$_3$ 3
(k) log 100   (l) $\log_2 8$
(m) antilog 3

(2.2) Without using a calculator or tables write down approximate values of

(a) $e^{0.16}$
(b) $\exp(-0.077)$

(2.3) (a) Rearrange the approximate formula you used in problem 2.2 to give an approximate formula for $\ln(1 + x)$ that applies when x

(positive or negative) is numerically small compared with 1. Then use it to write down approximate values of

- (b) $\ln 1.116$
- (c) $\ln 0.983$
- (d) $\ln 1.041$
- (e) $\ln 0.888$

(2.4) Given that $\ln 2 = 0.693$, $\ln 3 = 1.099$, $\ln 5 = 1.609$, $\ln 7 = 1.946$, evaluate the following:

- (a) $\ln 4$
- (b) $\ln 0.2$
- (c) $\ln 27$
- (d) $\ln 0.6$
- (e) $\ln 10$
- (f) $\exp(-0.693)$
- (g) $\exp(1.099)$
- (h) $\exp(2.708)$
- (i) $\exp(0.337)$

(2.5) The Boltzmann principle asserts that at thermal equilibrium the number of molecules in a state with energy E is proportional to $\exp(-E/RT)$, where $R = 8.31 \text{ J mol}^{-1}\text{K}^{-1}$ is the gas constant and T is the temperature in kelvins. In a proton magnetic resonance experiment any nucleus can exist in either of two spin states differing in energy (in a 100 MHz instrument) by about $4 \times 10^{-2} \text{ J mol}^{-1}$. Calculate the relative populations of nuclei in the two states.

(2.6) Sketch the form of a plot of mid-point potential against pH for a couple of the form

$$Ox + 2H^+ + 2\varepsilon^- \rightarrow RedH$$

assuming that $E^0 = +0.100 \text{ V}$ and RedH is a monobasic acid with $pK_a = 4.3$.

3 Differential Calculus

3.1 Co-ordinate geometry

Traditional geometry is concerned with the shapes of constructions on paper or in space and with the relationships between shapes. It has two distinct extensions into the more numerical domains of mathematics: *trigonometry* and *co-ordinate geometry*. Trigonometry is concerned with calculating the relationships between lengths and angles, whereas co-ordinate geometry has almost the reverse objective: it allows the use of geometrical insight and understanding for studying problems that are not essentially geometrical at all but algebraic. It happens that biochemistry is rather little concerned with real shapes, distances or angles; consequently neither traditional geometry nor trigonometry occupy the centre of the mathematical stage for the biochemist. Co-ordinate geometry, by contrast, is crucial, because many important relationships in physical chemistry and biochemistry appear in the first instance as algebraic expressions, and in this form they are too abstract to be immediately comprehended.

As an introduction to the use of co-ordinate geometry let us consider the following simple equation:

$$y = 7 + 3x$$

which expresses a relationship between two variables x and y. If we *plot* y against x as shown in Fig. 3.1, we are in effect using a geometrical model to 'map' an algebraic equation, giving concrete expression to what would otherwise be an abstraction. For example, if we let $x = 2$ the equation tells us that $y = 7 + 3 \times 2 = 13$, and we can plot the relationship by marking a point 2 units from the y axis in the x direction and 13 units from the x axis in the y direction. We can use the shorthand (2, 13), which simply means 'the point at which $x = 2$, $y = 13$', to refer to this point. It turns out that all points defined by the

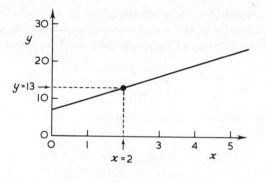

Fig. 3.1.   Plot of $y = 7 + 3x$.

equation lie on a straight line, as shown in Fig. 3.1, and indeed *any* equation of the form

$$y = a + bx$$

where a and b are constants, defines a straight line if y is plotted against x. This more general straight line is shown in Fig. 3.2.

It is obvious that $y = a$ if $x = 0$, because then the term bx is zero. Consequently the straight line cuts the y-axis a distance a from the point $(0, 0)$, the point where the axes intersect, which is known as the *origin*. Because of this relationship a is known as the *intercept* on the y-axis. The intercept on the x-axis is less obvious but may readily be found by putting $y = 0$: this gives $a + bx = 0$, and so $x = -a/b$, which is therefore the intercept on the x-axis.

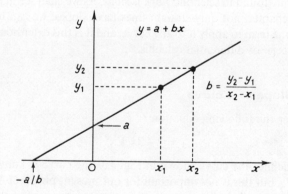

Fig. 3.2.   Plot of the general straight line $y = a + bx$.

The meaning of b in the general equation for a straight line follows from consideration of how y *changes* when x changes. Suppose that y changes from y_1 to y_2 as x changes from x_1 to x_2. Then

$$y_1 = a + bx_1$$

$$y_2 = a + bx_2$$

If the first equation is subtracted from the second the constant a disappears:

$$y_2 - y_1 = bx_2 - bx_1 = b(x_2 - x_1)$$

Thus

$$b = (y_2 - y_1)/(x_2 - x_1)$$

Notice that this relationship applies for *any* values of x_1 and x_2 we could have chosen (other than equal values, $x_2 = x_1$, which would introduce complexities that I prefer to avoid at present). In other words the ratio $(y_2 - y_1)/(x_2 - x_1)$ is a constant equal to b; indeed it is this constancy that defines algebraically what we mean when we say that a line is straight.

If b is numerically very small, the straight line defined by the equation $y = a + bx$ is almost parallel with the x-axis: y changes very slowly as x changes. Conversely, y changes very rapidly as x changes if b is numerically very large. Thus b has a meaning very similar to that of the *gradient* of a hill in everyday life. This term is sometimes used for the mathematical concept represented by b, but the term most commonly found in scientific work is *slope*. As we shall see in the rest of this chapter, not only straight lines have slopes: we can usefully extend the term to apply to any curve, and it is this extension that is the concern of differential calculus.

3.2 Slope of a curve

Consider the following equation:

$$y = 2 + 3x + x^2$$

which defines the curve shown in Fig. 3.3 (the curve is known as a *parabola*, but this is not important for our present purpose). We saw that for the straight line considered in the previous section the ratio $(y_2 - y_1)/(x_2 - x_1)$ was a constant. Let us now examine the same ratio

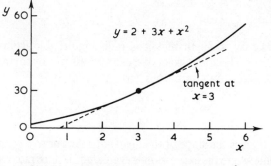

Fig. 3.3. Plot of the parabola $y = 2 + 3x + x^2$.

in the more complex example given in Fig. 3.3. First, it is convenient to introduce a new symbolism for $(y_2 - y_1)$ and $(x_2 - x_1)$. Let us use Δx to represent a change in x, and Δy to represent the *corresponding* (not just any) change in y. Then we can write $(y_2 - y_1)$ as Δy and $(x_2 - x_1)$ as Δx. So if

$$y = 2 + 3x + x^2$$

Then

$$y + \Delta y = 2 + 3(x + \Delta x) + (x + \Delta x)^2$$
$$= 2 + 3x + 3\Delta x + x^2 + 2x\Delta x + (\Delta x)^2$$

Subtracting, we have

$$\Delta y = 3\Delta x + 2x\Delta x + (\Delta x)^2$$

So

$$\frac{\Delta y}{\Delta x} = 3 + 2x + \Delta x$$

As we have said nothing about *how large* a change in x we are discussing, we have no grounds for simplifying this expression, but suppose we specify that Δx is a *small* change in x, i.e. small compared with $(3 + 2x)$, and let us emphasize this by replacing Δx with δx (a lower-case delta instead of a capital). Then in the similar expression

$$\frac{\delta y}{\delta x} = 3 + 2x + \delta x$$

we can justifiably make the approximation of ignoring δx in comparison with $(3 + 2x)$:

$$\frac{\delta y}{\delta x} \simeq 3 + 2x$$

If we make δx so small that it is indistinguishable from zero the approximation becomes exact. We can express this by saying that the *limit* of $\delta y/\delta x$ as δx approaches zero is $3 + 2x$, or

$$\underset{\delta x \to 0}{\text{Lim}} \frac{\delta y}{\delta x} = 3 + 2x$$

This result gives the slope of the *tangent* to the curve $y = 2 + 3x + x^2$ at the point where the expression is evaluated, as illustrated in Fig. 3.3. For convenience we often refer to this simply as the slope of the curve itself, although this is not a constant, unlike the slope of a straight line.

Now, it may be wondered why we have to refer to the 'limit as δx approaches zero' rather than the 'value when $\delta x = 0$', which may seem simpler and more natural. The difficulty with making $\delta x = 0$ is that then $\delta y/\partial x = 0/0$, a ratio that is generally undefined in mathematics because in principle it can have any value. We have just seen, however, that $\delta y/\delta x$ does not approach just any value as δx approaches zero but the exact and particular value $3 + 2x$. For this reason the more cumbersome terminology is a better description of the process.

Expressions such as $\underset{\delta x \to 0}{\text{Lim}} \dfrac{\delta y}{\delta x}$ are much too cumbersome for common use and so we *define* a new quantity $\dfrac{dy}{dx}$ as having exactly the same meaning, i.e. in general,

$$\frac{dy}{dx} \equiv \underset{\delta x \to 0}{\text{Lim}} \frac{\delta y}{\delta x}$$

This is called the *derivative* of y with respect to x; it can also be described as the result of *differentiating y* with respect to x. Quantities that are arbitrarily close to zero while remaining distinguishable from zero are called *infinitesimal*.

3.3 Rapid differentiation

The method we used in the previous section for differentiating $y = 2 + 3x + x^2$ is known as *differentiating from first principles*. It is quite general, i.e. the same approach can be used for differentiating any function. We shall use it again in this chapter to see how to obtain formulae for differentiating products, ratios and other functions. For everyday use it is rather long-winded, however, and in fact there is a simple rule that enables us to differentiate *any* power of x with respect to x. In general, if

$$y = Ax^i$$

where A and i are constants, then

$$\frac{dy}{dx} = Aix^{i-1}$$

Putting this into words, any constant multiplier remains unchanged, the index is decreased by 1 and the result is multiplied by the old index. Applying it to various examples, with A representing a constant in each case, we have

$$\frac{dA}{dx} = 0$$

$$\frac{d}{dx}(Ax) = A$$

$$\frac{d}{dx}(Ax^2) = 2Ax$$

$$\frac{d}{dx}(Ax^3) = 3Ax^2$$

$$\frac{d}{dx}(A/x) = \frac{d}{dx}(Ax^{-1}) = -Ax^{-2} = -A/x^2$$

$$\frac{d}{dx}(Ax^{\frac{1}{2}}) = \tfrac{1}{2}Ax^{-\frac{1}{2}} = A/2x^{\frac{1}{2}}$$

and so on. With the aid of the rules for differentiating combinations of terms that I shall discuss in the next three sections, this general rule permits the differentiation of almost all the expressions encountered in elementary biochemistry. Only two other simple cases are worth committing to memory, the derivatives of logarithms and exponentials:

$$\frac{d}{dx}(A\ln x) = A/x$$

$$\frac{d}{dx}(Ae^x) = Ae^x$$

3.4 Derivatives of sums and products

Suppose we wish to differentiate a sum, such as $y = u + v$, where u and v are both functions of x that can readily be differentiated with respect to x. By the arguments we have used before, a small change δx in x brings about changes in u and v as follows:

$$\delta u \simeq \frac{du}{dx}\delta x$$

$$\delta v \simeq \frac{dv}{dx}\delta x$$

Clearly if $y = u + v$ the change in y must be the sum of those in u and v:

$$\delta y = \delta u + \delta v$$

So

$$\frac{\delta y}{\delta x} \simeq \frac{du}{dx} + \frac{dv}{dx}$$

In the limit the approximation becomes exact, i.e.

$$\frac{dy}{dx} = \underset{\delta x \to 0}{\text{Lim}} \frac{\delta y}{\delta x} = \frac{du}{dx} + \frac{dv}{dx}$$

This result can be extended to the sum of three or more functions of x, and in general *the derivative of a sum is equal to the sum of the derivatives*. The derivative of a difference follows directly:

$$\frac{d}{dx}(u - v) = \frac{du}{dx} - \frac{dv}{dx}$$

as a difference is simply a sum in which the second term is negative.

Suppose now that $y = uv$ is the *product* of two functions of x. Then

$$y + \delta y = (u + \delta u)(v + \delta v)$$

$$= uv + u\delta v + v\delta u + \delta u\delta v$$

so, by subtraction,

$$\delta y = u\delta v + v\delta u + \delta u\delta v$$

and, dividing every term by δx:

$$\frac{\delta y}{\delta x} = u\frac{\delta v}{\delta x} + v\frac{\delta u}{\delta x} + \frac{\delta u\delta v}{\delta x}$$

In the limit the last term disappears, because it is itself a small quantity whereas the other terms in the expression are simply ratios of small quantities:

$$\frac{dy}{dx} = \underset{\delta x \to 0}{\text{Lim}} \frac{\delta y}{\delta x} = u\frac{dv}{dx} + v\frac{du}{dx}$$

This derivation is illustrated graphically in Fig. 3.4: the change in uv is represented by the sum of the areas in the three rectangles labelled

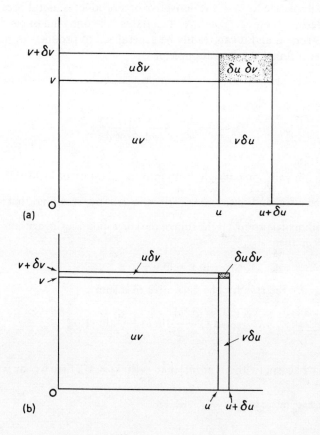

Fig. 3.4.   Explanation of the formula for the derivative of a product uv. In (a) the increments δv and δu are appreciable and the shaded area is by no means negligible, but as the increments become smaller, as in (b), the shaded area decreases very rapidly, not only absolutely, but also in relation to the areas labelled $u\delta v$ and $v\delta u$, which become smaller much more slowly.

$u\delta v$, $v\delta u$ and $\delta u\delta v$. It is clear from inspection that as δu and δv become small, $\delta u\delta v$ becomes smaller very much faster, and in the limit can be ignored.

If we divide the left-hand side of the above equation by y and the right-hand side by uv ($= y$) we have

$$\frac{1}{y}\frac{dy}{dx} = \frac{1}{u}\frac{du}{dx} + \frac{1}{v}\frac{dv}{dx}$$

This alternative form of the derivative of a product is useful because each term is 'homogeneous', i.e. it contains only one kind of variable apart from x, and it can readily be generalized to products of more than two functions. For example, if

$$y = uvw$$

then

$$\frac{1}{y}\frac{dy}{dx} = \frac{1}{u}\frac{du}{dx} + \frac{1}{v}\frac{dv}{dx} + \frac{1}{w}\frac{dw}{dx}$$

and so on.

We can gain some insight both into the ubiquity of logarithms in mathematics and into *why* $\frac{d}{dx}(\ln x) = 1/x$ by recalling that the logarithm of a product is the sum of the corresponding logarithms, i.e. if $y = uv$ then

$$\ln y = \ln u + \ln v$$

and so, by the rule for the derivative of a sum,

$$\frac{d}{dx}(\ln y) = \frac{d}{dx}(\ln u) + \frac{d}{dx}(\ln v)$$

but if we accept (without proof) that $\frac{d}{dx}(\ln x) = 1/x$ then we can write $d \ln x$ as $\frac{1}{x}dx$, $d \ln y$ as $\frac{1}{y}dy$, etc., so

$$\frac{1}{y}\frac{dy}{dx} = \frac{1}{u}\frac{du}{dx} + \frac{1}{v}\frac{dv}{dx}$$

which is the same result as we had before for the derivative of a product. Although this does not prove that our expression for the derivative of ln x is correct it does show that it is consistent with what we know independently and hence that it is plausible.

3.5   Derivative of a 'function of a function'

The bell-shaped curves often found for the pH dependence of protein and enzyme properties (such as catalytic activity) can be represented by an equation of the following sort, known as a *Michaelis function*:

$$y = \frac{A}{1 + (B/x) + Cx} = A(1 + Bx^{-1} + Cx)^{-1}$$

where y is some property of the protein, x is (usually) the hydrogen-ion concentration, i.e. not pH but 10^{-pH}, and A, B and C are constants. If we write the denominator as

$$u = 1 + Bx^{-1} + Cx$$

then the expression as a whole can be written as

$$y = Au^{-1}$$

So it is easy to differentiate u with respect to x:

$$\frac{du}{dx} = -Bx^{-2} + C$$

and it is easy to differentiate y with respect to u:

$$\frac{dy}{du} = -Au^{-2}$$

But how can we combine the results to get the derivative of y with respect to x? This is an example of the problem of how to differentiate a *function of a function* – y is a function of u, which is itself a function of x. I shall now consider this as a general problem before returning to the specific example of the Michaelis function.

For $y = F(u)$, $u = f(x)$, we have

$$\delta y \simeq \frac{dy}{du}\delta u, \; \delta u \simeq \frac{du}{dx}\delta x$$

and so

$$\delta y \simeq \frac{dy}{du}\frac{du}{dx}\delta x$$

and

$$\frac{\delta y}{\delta x} \simeq \frac{dy}{du}\frac{du}{dx}$$

and in the limit

$$\frac{dy}{dx} = \lim_{\delta x \to 0} \frac{\delta y}{\delta x} = \frac{dy}{du}\frac{du}{dx}$$

In general, if f_1 is a function of f_2, f_2 is a function of f_3, etc., to f_n, which is a function of x, then

$$\frac{df_1}{dx} = \frac{df_1}{df_2}\frac{df_2}{df_3}\frac{df_3}{df_4} \cdots \frac{df_n}{dx}$$

This is known as the *chain rule*. It may seem obvious, as it can be 'derived' by treating all of the derivatives as fractions and cancelling the like elements. Although such a 'derivation' would not satisfy a serious mathematician the differences between derivatives and fractions rarely create difficulties in biochemistry. Consequently it does little harm if the biochemist uses non-rigorous 'derivations' as an aid to remembering important general relationships.

Returning to the Michaelis function, application of the chain rule shows the derivative to be

$$\frac{dy}{dx} = \frac{dy}{du}\frac{du}{dx} = -Au^{-2}(-Bx^{-2} + C) = \frac{ABx^{-2} - AC}{(1 + Bx^{-1} + Cx)^2}$$

3.6 Derivative of a ratio

A ratio can be regarded as a product in which the second term is raised to the power -1. So $y = u/v$ is the same as $y = uv^{-1}$, and

$$\frac{dy}{dx} = u\frac{d}{dx}(v^{-1}) + v^{-1}\frac{du}{dx}$$

in which v^{-1} is a function of a function, with derivative

$$\frac{d}{dx}(v^{-1}) = \frac{d}{dv}(v^{-1})\frac{dv}{dx} = -v^{-2}\frac{dv}{dx}$$

So

$$\frac{dy}{dx} = -uv^{-2}\frac{dv}{dx} + v^{-1}\frac{du}{dx} = \frac{v\dfrac{du}{dx} - u\dfrac{dv}{dx}}{v^2}$$

3.7 Higher derivatives

After a function has been differentiated once there is no reason why the result should not be differentiated a second time by application of exactly the same rules. For example,

$$y = 7 + 3x - 2x^2 + x^3$$

$$\frac{dy}{dx} = 3 - 4x + 3x^2$$

$$\frac{d}{dx}\left(\frac{dy}{dx}\right) = -4 + 6x$$

This last quantity is usually written more compactly as d^2y/dx^2 (spoken aloud as 'dee two y by dee x squared' *not* as 'dee squared y by dee x squared') and is called the *second derivative* of y with respect to x. Just as the first derivative expresses how fast y changes with x, the second derivative expresses how fast dy/dx changes with x. Alternatively, if dy/dx expresses the *slope* of a plot of y against x, d^2y/dx^2 expresses the rate of change of slope or in other words the *curvature*. If there is no curvature the slope is constant, so the second derivative is zero. For a general straight line, therefore,

$$y = A + Bx$$

$$\frac{dy}{dx} = B$$

$$\frac{d^2y}{dx^2} = 0$$

3.8 Notation

The notation that I have used hitherto in this chapter is based on that of the great German mathematician Leibniz. Although it is the most generally useful and the most widely used there are circumstances in which it becomes cumbersome, and one should have some familiarity with three other systems that are sometimes used.

(1) In mathematical tables, which one may wish to consult to obtain unfamiliar results, it is usually obvious what variable one is differentiating with respect to, and compactness is achieved by writing d/dx as D. Many people also find this a useful form for memorizing the standard results for the derivatives of products and ratios:

$$D(uv) = uDv + vDu$$

$$D(u/v) = (vDu - uDv)/v^2$$

(2) Another system for achieving compactness when it is obvious what one is differentiating with respect to is to write the first derivative of y with respect to x as y', and the second derivative as y''. This is especially convenient when one does not want to define a particular variable (y in this example) as the function, e.g. one can write the first derivative of $f(x)$ as $f'(x)$, and the second as $f''(x)$.

(3) Calculus was developed independently by Newton in England and Leibniz in Germany. Newton's *dot notation* is in general much less convenient than that of Leibniz, partly because it does not allow for the possibility that one may wish to differentiate with respect to more than one variable. Newton was most concerned with functions of *time*, or t, and when dealing with time (e.g. in enzyme kinetics) it is still occasionally convenient to write dy/dt as $\dot{y}$ and d^2y/dt^2 as $\ddot{y}$. Even in kinetics, however, this notation is much less common than that of Leibniz.

The various alternative systems are listed in Table 3.1. As indicated in the table, one can continue differentiating beyond the second derivative, but the biochemist rarely wants to do this and so I have not

Table 3.1 Notations for expressing derivatives

Function	y	y	y	$f(x)$	$x = f(t)$
1st derivative	$\dfrac{dy}{dx}$	Dy	y'	$f'(x)$	$\dot{x}$
2nd derivative	$\dfrac{d^2y}{dx^2}$	D^2y	y''	$f''(x)$	$\ddot{x}$
nth derivative	$\dfrac{d^ny}{dx^n}$	D^ny	$y^{(n)}$	$f^{(n)}(x)$	—

discussed it in the text. It will be noticed that although Leibniz's notation is the least compact it is also the most explicit. For this reason it is nearly always wise for the non-mathematician to prefer it: a small saving in space is never justified if it leads to bafflement.

3.9 Maxima and minima

Consider the following equation:

$$v = \frac{Vs}{K_m + s + s^2/K_{si}}$$

which represents a simple kind of *substrate inhibition* in an enzyme-catalysed reaction. The variables are the rate, v, and the concentration of substrate, s, and V, K_m and K_{si} are constants. Unlike the examples earlier in this chapter, this equation is not written with the typical symbols of elementary mathematics, x, y, A, B, etc., but with the sort of symbols most likely to be encountered in biochemical work. This difference should cause no problems: one does not require a particular set of symbols to carry out mathematical operations, and it is unwise to associate particular relationships too rigidly with particular symbols.

Fig. 3.5 shows the form of a plot of v against s for the equation for

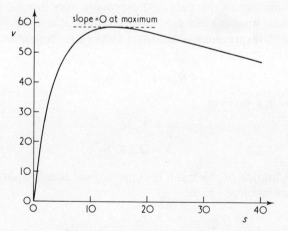

Fig .3.5. Plot of v against s for a system showing substrate inhibition. The plot illustrates the point that we can define a maximum as a point at which the slope is zero, but this definition would be incomplete because it would also include minima (see Fig. 3.6).

substrate inhibition. A striking feature of the curve is that instead of increasing indefinitely with s (as in simpler kinds of kinetic behaviour) v increases to a *maximum* and then decreases as s is increased further. A major use of the differential calculus is that it enables one to calculate where a maximum will occur. From inspection of the graph it can be seen that at the maximum the slope is zero. This is true in general, i.e. at *any* maximum the slope is zero (although the converse is not, because a point of zero slope may also represent a minimum). Accordingly, we can find the maximum in Fig. 3.5 by differentiating v with respect to s and finding a value of s at which $\mathrm{d}v/\mathrm{d}s$ is zero. This can be done by applying the rule for differentiating a ratio:

$$\begin{aligned}
\frac{\mathrm{d}v}{\mathrm{d}s} &= \frac{(K_m + s + s^2/K_{si})V - Vs(1 + 2s/K_{si})}{(K_m + s + s^2/K_{si})^2} \\
&= \frac{VK_m + Vs + Vs^2/K_{si} - Vs - 2Vs^2/K_{si}}{(K_m + s + s^2/K_{si})^2} \\
&= \frac{VK_m - Vs^2/K_{si}}{(K_m + s + s^2/K_{si})^2}
\end{aligned}$$

In principle, an expression such as this can be zero either because the numerator is zero or because the denominator is infinite. Although the denominator of this particular expression does increase steeply and without limit as s increases it remains finite at all finite s. So the only way the expression can be zero at finite s is for the numerator to be zero, i.e.

$$VK_m - Vs^2/K_{si} = 0$$

Solving for s, we have

$$s^2 = K_m K_{si}$$

$$s = (K_m K_{si})^{\frac{1}{2}}$$

and substitution of this result into the original equation shows the maximum value of v to be

$$v = \frac{V(K_m K_{si})^{\frac{1}{2}}}{K_m + (K_m K_{si})^{\frac{1}{2}} + K_m K_{si}/K_{si}}$$

$$= \frac{V(K_m K_{si})^{\frac{1}{2}}}{2K_m + (K_m K_{si})^{\frac{1}{2}}} = \frac{V}{1 + 2(K_m/K_{si})^{\frac{1}{2}}}$$

The Michaelis function that we considered earlier in this chapter is of exactly the same form as this equation for substrate inhibition, because if

$$y = \frac{A}{1 + (B/x) + Cx}$$

then multiplication of both numerator and denominator by x gives

$$y = \frac{Ax}{B + x + Cx^2}$$

which is exactly the same equation with different symbols (y for v, x for s, A for V, B for K_m and $1/C$ for K_{si}). Consequently the maximum of the Michaelis function must occur when $x = (B/C)^{\frac{1}{2}}$, $y = A/[1 + 2(BC)^{\frac{1}{2}}]$. In this case, however, we are more likely to want to plot y against $\log x$ than against x (because the Michaelis function is a dependence on hydrogen-ion concentration, which is usually expressed in logarithmic form). The question therefore arises, can we assume that the same maximum occurs in a plot of y against $\log x$? Put differently, can we assume that $dy/d\log x = 0$ when $dy/dx = 0$? The answer comes from considering the relationship between the two derivatives, which follows from the chain rule:

$$\frac{dy}{d\log x} = \frac{dy}{dx}\frac{dx}{d\ln x}\frac{d\ln x}{d\log x} = 2.303x\frac{dy}{dx}$$

(because $d\ln x/dx = 1/x$ and $\log x = 2.303 \ln x$ approximately). Thus, provided x is finite, a zero value of dy/dx must occur at the same value of x as a zero value of $dy/d\log x$, and so the maximum in the plot of y against $\log x$ occurs at the same values of x and y as in the plot of y against x.

Exactly the same criteria can be used to find a *minimum* of a function, because this also corresponds to a zero value of the first derivative (Fig. 3.6). Clearly, therefore, the existence of such a zero value does not guarantee that a maximum has been found. It shows only that one has found a *stationary point*, a more general term that embraces both maxima and minima. In most elementary biochemical circumstances this is not a problem because it is usually obvious whether one is dealing with a maximum or a minimum. In the more general case there may be ambiguity, but this can be resolved by examining the sign of slope on either side of the stationary point. At a

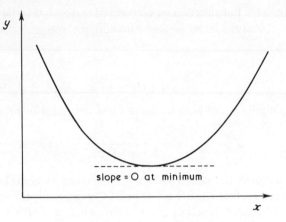

Fig. 3.6.   Minimum of an arbitrary curve. Again, the slope is zero at the minimum (see Fig. 3.5), but it is increasing through zero as x increases.

maximum, the slope is *decreasing* through zero, whereas at a minimum it is *increasing* through zero. Returning to the expression for dv/ds in the case of substrate inhibition, it is clear that the numerator, $VK_m - Vs^2/K_{si}$, must decrease continuously as s is increased through the positive range, and so it must decrease through zero as s increases through the value $(K_m K_{si})^{\frac{1}{2}}$. This is, therefore, a maximum, as assumed previously, not a minimum. It is not necessary to consider the denominator of the expression for dv/ds because this contains positive terms only (for positive s) and must therefore be positive.

3.10   A note on terminology

It is often true that terms in mathematics have exact meanings that do not precisely agree with the meanings of the same words when used in everyday language. An example is the term 'maximum', which refers in mathematics, as we have seen, to a point at which the first derivative of a function is zero and is decreasing as the abscissa variable increases. For a complex function there may be several or many such points, and in such a case we would refer to them as 'maxima' and to one of them as 'a maximum', not as 'the maximum'. Similar considerations apply to minima. Moreover, there is nothing in the definitions of either that requires a maximum to occur at a larger value of the function than a minimum; indeed one can find functions that have a unique maximum at a lower value than the unique minimum, such as

$$y = x + \frac{1}{x}$$

(If you find this statement puzzling you should plot y as given by this equation against x, for x between -2 and -0.1 and between 0.1 and 2.) This sort of result would be absurd in everyday language, in which *the* maximum of a measurement is *the* largest value it can have and *the* minimum is *the* smallest value it can have. Moreover, 'it can have' is usually taken to mean 'under realistic circumstances'.

Because biochemistry was not developed largely by mathematicians, mathematical terms have come to be used in biochemistry with looser meanings than would be acceptable to mathematicians. For example, the quantity V in the Michaelis–Menten equation:

$$v = Vs/(K_m + s)$$

is called the *maximum velocity*, because v cannot exceed it. However, v cannot attain the value V at a finite value of s and thus V is not a maximum in the mathematical sense but a limit. Moreover, when a true maximum does occur in a plot of v against s, as in the example of substrate inhibition discussed in the previous section, the value of v at the maximum should not be called the 'maximum velocity', because this would cause confusion with the usual meaning of the term as V.

3.11 Points of inflexion

An alternative to the method I have described for deciding whether a stationary point is a maximum or a minimum is to examine the sign of the second derivative at the stationary point. This is because the second derivative expresses directly how the first derivative is changing with the abscissa variable. At a maximum the first derivative is decreasing and so the second derivative is normally negative; conversely, at a minimum it is normally positive. Although I could illustrate this with the same example as in Section 3.9 we would then get a rather complicated expression on differentiating twice; it is easier to consider a function such as the following:

$$y = 2 + 9x - 6x^2 + x^3$$

In this case

$$\frac{dy}{dx} = 9 - 12x + 3x^2$$

$$\frac{d^2y}{dx^2} = -12 + 6x$$

Setting $dy/dx = 0$ gives a quadratic equation with the two solutions $x = 1$, $y = 6$ and $x = 3$, $y = 2$. As 6 is larger than 2, it may seem obvious that the first solution is a maximum and the second a minimum. However, although this conclusion happens to be correct in this instance it does not follow from the logic, which is unsound (recall the function $y = x + \dfrac{1}{x}$ mentioned in the previous section). It is safer therefore to examine the value of d^2y/dx^2: this is -6 at $x = 1$ and $+6$ at $x = 3$, and shows that the point $(1, 6)$ is indeed a maximum and the point $(3, 2)$ a minimum.

Occasionally dy/dx and d^2y/dx^2 are both zero at the same value of x, but this is unusual in the sort of equations that occur in biochemistry and thus rarely interferes with the method just described for distinguishing between maxima and minima. It is, however, important to consider zero values of d^2y/dx^2 that do not necessarily coincide with zero values of dy/dx. These occur at so-called *points of inflexion*, which are points at which the slope of the plot is a maximum or a minimum (Fig. 3.7).

Points of inflexion are important in biochemistry because they define conditions in which a response (e.g. the rate of a reaction) is most (or least) sensitive to an influence (e.g. the concentration of a metabolite). Consider, for example, the ability of a buffer to resist the changes in pH that might be brought about by addition of alkali. The equation that defines this (to a first approximation) is the *Henderson–Hasselbalch equation*:

$$pH = pK_a + \log \frac{[\text{salt}]}{[\text{acid}]}$$

For example, for a buffer made up by adding x mol l^{-1} NaOH to a solution of acetic acid, HOAc, at an initial concentration A, we have:

$$pH = pK_a + \log \frac{[\text{NaOAc}]}{[\text{HOAc}]}$$

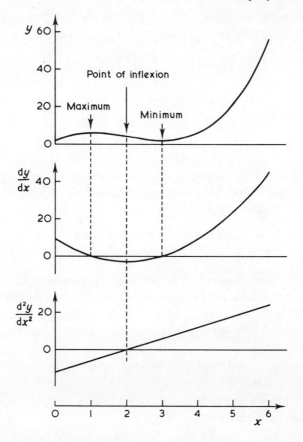

Fig. 3.7. Point of inflexion. At a maximum or minimum the first derivative is zero; at a point of inflexion the first derivative is a maximum or a minimum and the second derivative is zero.

$$= pK_a + 0.4343 \ln \frac{[\text{NaOAc}]}{[\text{HOAc}]}$$

$$= pK_a + 0.4343 \ln \frac{x}{A - x}$$

because as long as x is less than A essentially all of the added OH^- ions convert HOAc molecules stoichiometrically into OAc^- ions.

(0.4343 is the reciprocal of 2.303, or $1/\ln 10$). Differentiating with respect to x, we have

$$\frac{\mathrm{dpH}}{\mathrm{d}x} = 0.4343 \left(\frac{A-x}{x}\right) \left[\frac{(A-x)-x(-1)}{(A-x)^2}\right]$$

$$= \frac{0.4343\,A}{x(A-x)}$$

This first derivative is a measure of the sensitivity of the pH to addition of base. Clearly the buffer is most effective when it is a minimum (or its reciprocal, known as the *buffer capacity*, is a maximum). To find the value of x that makes $\mathrm{dpH}/\mathrm{d}x$ a minimum we must differentiate again:

$$\frac{\mathrm{d}^2\mathrm{pH}}{\mathrm{d}x^2} = \frac{-0.4343\,A(A-2x)}{(Ax-x^2)^2}$$

By inspection it is clear that this derivative is zero when $x = \frac{1}{2}A$, i.e. when enough base has been added to neutralize exactly half of the original acid. Substituting back into the Henderson–Hasselbalch equation, we find that

$$\mathrm{pH} = \mathrm{p}K_a + 0.4343 \ln\left(\frac{2A}{2A}\right) = \mathrm{p}K_a$$

Thus a buffer is most effective at the pH equal to the $\mathrm{p}K_a$ of the acid involved in the titration.

3.12 Sketching curves

One of the most important uses of differential calculus is as an aid to sketching the curves corresponding to unfamiliar functions. Suppose, for example that one had found that the concentration q of product of a reaction at time t could be expressed by an equation of the following form, in which all symbols apart from q and t are positive constants:

$$q = \frac{k_{+2}k_{+3}e_0 t}{k_{+2}+k_{+3}} - \frac{k_{+2}k_{+3}e_0\{1-\exp[-(k_{+2}+k_{+3})t]\}}{(k_{+2}+k_{+3})^2}$$

It is by no means obvious at first glance what sort of curve this equation defines, but one can deduce a good deal simply by examining two extreme values of t. First, if $t = 0$ then

$$q = 0 - \frac{k_{+2}k_{+3}e_0[1 - \exp(0)]}{(k_{+2} + k_{+3})^2} = 0$$

So the curve passes through the origin. Second, if t is large enough for $\exp[-(k_{+2} + k_{+3})t]$ to be negligible compared with 1, the equation simplifies to

$$q = \frac{k_{+2}k_{+3}e_0 t}{k_{+2} + k_{+3}} - \frac{k_{+2}k_{+3}e_0}{(k_{+2} + k_{+3})^2}$$

This is the expression of a straight line with a positive slope and a negative intercept on the ordinate, and so the curve must approximate to such a straight line at large values of t.

Although this information is useful it tells us little about what might well prove to be the most interesting part of the time course, the period before the straight-line approximation becomes valid. Light can be shed on this period by differentiating twice to reveal the

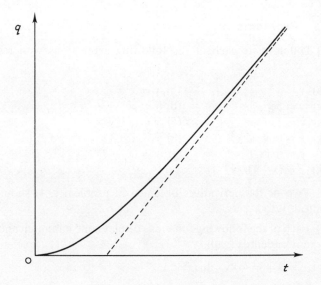

Fig. 3.8.   Sketch of a complex curve (defined in the text) obtained by considering the value of q and its first and second derivatives with respect to t at various values of t. The broken line is a straight line and represents the limit approached by the curve as t increases.

behaviour of the slope:

$$\frac{dq}{dt} = \frac{k_{+2}k_{+3}e_0}{k_{+2}+k_{+3}} - \frac{k_{+2}k_{+3}e_0\exp[-(k_{+2}+k_{+3})t]}{k_{+2}+k_{+3}}$$

$$= \frac{k_{+2}k_{+3}e_0\{1-\exp[-(k_{+2}+k_{+3})t]\}}{k_{+2}+k_{+3}}$$

$$\frac{d^2q}{dt^2} = k_{+2}k_{+3}e_0\exp[-(k_{+2}+k_{+3})t]$$

As $\exp[-(k_{+2}+k_{+3})t]$ must have a value between 1 (at $t = 0$) and 0 (for $t \to \infty$) for any positive value of t, it follows that dq/dt cannot be negative and is zero only at the origin. Thus the plot of q against t cannot contain a maximum or a minimum except at the origin. Moreover d^2q/dt^2 is positive at all values of t, although its value becomes negligible at large values of t. When all of this information is combined it becomes clear that the form of the curve must be as shown in Fig. 3.8.

3.13   Problems

(3.1) Differentiate each of the following expressions with respect to x:

(a) $y = x^3$ (b) $y = x^{1/2}$
(c) $y = 5e^x$ (d) $y = e^{5x}$
(e) $y = 3\ln x$ (f) $y = x^{1/2} + x^{-1/2}$
(g) $y = \ln x - 2x^2$ (h) $y = x^2e^x$
(i) $y = \ln(x^3)$ (j) $y = (x^2+3)^{1/2}$
(k) $y = (x+1)/(x-1)$

(3.2) Two of the derivatives obtained in problem (3.1) should be identical. Why?

(3.3) Each of the following functions displays one minimum and one maximum. Identify both:

(a) $y = 7 + x^2 - 8x + \ln(x^6)$

(b) $y = 15 - x - 1/(x-5)$

(3.4) Michaelis and Menten analysed their data in terms of the equation that bears their names, $v = Vs/(K_m + s)$, by plotting v against $\log s$. They used the facts that this plot has a point of inflexion at

$s = K_m$ and that the maximum slope is $0.576V$. Prove that these facts are correct.

(3.5) Differentiate the equation $v = Vs/(K_m + s)$ with respect to s. What are the values of dv/ds when (a) $s = 0$; (b) $s = K_m$; (c) s approaches infinity?

(3.6) Consider a pH profile defined by the following equation, in which h represents the hydrogen-ion concentration $[H^+]$ and $\hat{y}$ is a constant:

$$y = \frac{y}{1 + (h/K_1) + (K_2/h)}$$

(a) At what value of h is y a maximum?
(b) Show that this solution is equivalent to $pH = (pK_1 + pK_2)/2$.

(3.7) Under certain limiting conditions the binding of a small ligand to a protein can be approximated by an equation of the following form:

$$Y = \frac{Kx^h}{1 + Kx^h}$$

in which Y is a measure of the extent of binding, x is the concentration of ligand, and K and h are constants. Putting $K = 1$ (arbitrary unit), $h = 2$, sketch the form of curve given by this equation for a plot of Y against x (for positive x).

(3.8) For the equation given in problem (3.7), what would be the slope of a plot of $\log[Y/(1 - Y)]$ against $\log x$?

(3.9) Differentiate the following equation with respect to t, treating p and t as variables and V, K_m and s_0 as constants:

$$Vt = p + K_m \ln[s_0/(s_0 - p)]$$

4 Integral Calculus

4.1 Increases in area

Fig. 4.1 represents an arbitrary function y of x, and the area of the region bounded by the curve, the x-axis and the verticals $x = x_0$ and $x = x_1$ is designated A. What is the increase δA in this area if x_1 is increased to $x_1 + \delta x$? Apart from the small approximately triangular area that is shaded in the diagram, all of the increase in area is accounted for by the tall thin rectangle of height y_1 and width δx, which has area $y_1 \delta x$. In symbols,

$$\delta A \simeq y_1 \delta x$$

If x is made smaller, the small shaded triangle decreases rapidly in area, not only absolutely, but also in relation to the area of the rectangle. Consequently, the approximation embodied by the above equation is very accurate if δx is very small, and in the limit as δx approaches zero it becomes exact. This is not very helpful, however, if we want to know the value of δA when δx is large! We can overcome the difficulty by noting that if we divide both sides of the equation by δx we have

$$\frac{\delta A}{\delta x} \simeq y$$

which in the limit is exact:

$$\frac{\mathrm{d}A}{\mathrm{d}x} = y$$

This expresses the important generalization that the *derivative of the area between the x-axis and any curve $y = f(x)$ is equal to the ordinate variable y*. As we shall see, this relationship means that we can

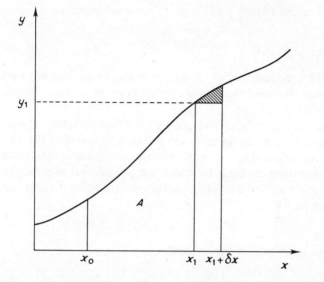

Fig. 4.1.   Area under a curve. Ignoring the shaded area, the increase in A as x increases from x_1 to $x_1 + \delta x$ is $y_1 \delta x$.

calculate the area under any curve whose algebraic expression can be recognized as the derivative of some other function. First, however, we must see how a large area can be treated as the sum of many small areas.

As long as δx is small enough for the relationship $\delta A \simeq y \delta x$ to be a good approximation we can write

$$\Delta A \simeq \sum_{}^{n} \delta A = \sum_{}^{n} y \delta x$$

for the increase in area ΔA brought about by increasing x by $\Delta x = n \delta x$. At first sight this may seem simple to evaluate, as $\sum_{}^{n} \delta x$ is just $n \delta x$, or Δx; but to get a good approximation we must not treat y as a constant but must constantly recalculate it as x changes. If the approximation is made better and better by making δx smaller and smaller the number of increments n must be increased concomitantly to keep Δx the same size. In the limit the relationship is exact when there are an infinite number of infinitesimal increments:

$$\Delta A = \lim_{n \to \infty} \sum^{n} y\delta x = \int y\mathrm{d}x$$

in which the new symbol $\int$ is used to represent this sum of an infinite number of terms. We call the expression an *integral* and the process *integration*.

Because of the relationship $\mathrm{d}A/\mathrm{d}x = y$ that we referred to earlier, we can in principle integrate any function by reversing the rules for differentiation. (In practice this is often more difficult than it sounds, because there are many functions that are difficult or impossible to recognize as the results of differentiating other functions.) For example, if

$$y = 3x^2 + 2x + 5$$

then

$$\int y\mathrm{d}x = x^3 + x^2 + 5x + \alpha$$

where α is a constant, because differentiation of this second function with respect to x gives the first function.

4.2   Definite and indefinite integrals

We have seen that integration can be regarded as the inverse of differentiation. Just as the derivative of any constant is zero, therefore, the integral of zero is any constant and the constant α that appeared in the integration at the end of the previous section can have any value. A constant of this sort is called a *constant of integration*. Its meaning, and the reason for its appearance, can be seen by reflecting that in the previous section we discussed the *increase* in an area A without specifying what value of A we had to start off with.

The ambiguity implied by the need for a constant of integration can be resolved by specifying *limits* between which the integration is to be done. Thus, an expression such as

$$\int (3x^2 + 2x + 5)\mathrm{d}x$$

is ambiguous because although it tells us to add together terms of the form $(3x^2 + 2x + 5)\mathrm{d}x$ it does not tell us where to begin and end. This sort of integral is called an *indefinite integral*, but if we wish to start at $x = 1$ and finish at $x = 2$ we can indicate this by writing the integral as a *definite integral*:

$$\int_1^2 (3x^2 + 2x + 5)dx$$

This can be regarded as the difference between the values of the indefinite integral at $x = 1$ and at $x = 2$:

$$\int (3x^2 + 2x + 5)dx = x^3 + x^2 + 5x + \alpha$$
$$= 7 + \alpha \text{ at } x = 1, 22 + \alpha \text{ at } x = 2$$

On subtracting one from the other the ambiguity disappears because regardless of the value of α it vanishes from the difference:

$$\int_1^2 (3x^2 + 2x + 5)dx = \left[x^3 + x^2 + 5x\right]_1^2 = 22 - 7 = 15$$

It is sometimes convenient, as in this example, to indicate the limits outside square brackets after the form of the indefinite integral has been decided. An expression of this sort can be read as an instruction to 'evaluate the expression contained in the bracket at both limits and subtract one from the other'.

Notice that an indefinite integral is a *function* that can have any value, not only because of the undefined constant, but also because it can be evaluated at any value of x. By contrast, a definite integral defines a *number*. It follows that evaluation of a definite integral requires two stages: first, determination of the function corresponding to the indefinite integral; and second, evaluation of this function at the specified limits. In general, the second stage is much easier than the first: given an indefinite integral we can always evaluate it, but it is not unusual to begin with a function that we cannot integrate at all.

As a chemical example, consider the rate of a simple first-order reaction, which can be expressed as

$$\frac{da}{dt} = -ka$$

where a is the concentration of the reacting substance at time t and k is a constant (the minus sign is to show that the reacting substance disappears, i.e. its concentration decreases, as the reaction proceeds). We can rearrange this to show that an infinitesimal change da in a is related to an infinitesimal change dt in t as follows:

$$\frac{da}{a} = -kdt$$

Suppose we want to know the total decrease in concentration that occurs between $t = 0$ and $t = t_1$: we can find this by integrating:

$$\int_{t=0}^{t=t_1} \frac{da}{a} = -k \int_0^{t_1} dt$$

In this case the indefinite integrals are straightforward because, on the left-hand side, we know that $\frac{d}{dx}(\ln x) = \frac{1}{x}$ and, on the right-hand side, that $\frac{d}{dx}(x) = 1$. So

$$\left[\ln a\right]_{t=0}^{t=t_1} = -k\left[t\right]_0^{t_1}$$

If we define $a = a_0$ at $t = 0$ and $a = a_1$ at $t = t_1$, this becomes

$$\ln a_1 - \ln a_0 = -kt_1$$

which can be rearranged to

$$a_1 = a_0 \exp(-kt_1)$$

and hence specifies the concentration a at any time.

In this derivation I have been pedantic in two respects: first, where the limits are expressed in terms of a different variable from the one that appears in the function to be integrated, as on the left-hand side of the equation above, we should express them as '$t = 0$', etc., rather than simply as '0', which, in the above example, would imply that a was 0 at the lower limit. Second, t_1 and t were given different symbols because they were used with different meanings – t was a variable whereas t_1, a constant, was a particular value of t. This distinction is almost universally ignored in modern chemical practice, i.e. the result above would normally be written as

$$a = a_0 \exp(-kt)$$

This practice does no harm provided one realizes what has been done and one is willing to tolerate expressions such as $\int_0^t dt$ in which the same symbol t is used with two different meanings. It is especially convenient in applications where one does want to treat the upper limit as a variable in subsequent analysis.

In kinetic applications it is often clearer not to write definite integrals at all but to work with indefinite integrals with constants of integration that can be defined or evaluated. Therefore we might write the initial equation for the first-order reaction as

$$\int \frac{\mathrm{d}a}{a} = -k \int \mathrm{d}t$$

and integrate it to give

$$\ln a = -kt + \alpha$$

Notice that the same constant of integration α serves both integrals: if we had included constants of integration on both sides we could subtract one from the other to get the constant we have called α. If we define $a = a_0$ at $t = 0$, we have

$$\ln a_0 = \alpha$$

and so

$$\ln a = -kt + \ln a_0$$

and rearranging gives us the same result as before,

$$a = a_0 \exp(-kt)$$

but without any confusion about whether a and t are variables or constants.

It is important to note that the constant of integration in this example was not zero, even though it was evaluated at $t = 0$. With a few elementary kinds of function–most notably polynomials, i.e. $A + Bx + Cx^2 + Dx^3 + \ldots$–the integral has a value of zero when the lower limit is zero. However, although these functions are common in elementary textbook accounts of integration they are not common in scientific applications of integration and one must not suppose that there is any kind of general rule that allows constants of integration to be ignored. In kinetics, which accounts for most of the integrating that a biochemist is likely to do, it is never safe to ignore constants of integration or to assume that integrals are zero at zero time.

4.3 Simple integrals

In principle, any function can be differentiated, but the converse is by
no means true. The only functions that are easy to integrate are those
that are recognizable as derivatives of known functions. For example,

$$\int A\,\mathrm{d}x = Ax + \alpha$$

$$\int Ax\,\mathrm{d}x = \tfrac{1}{2}Ax^2 + \alpha$$

$$\int Ax^2\,\mathrm{d}x = \tfrac{1}{3}Ax^3 + \alpha$$

or in general,

$$\int Ax^i\,\mathrm{d}x = \frac{Ax^{i+1}}{i+1} + \alpha$$

The rule for terms of polynomials is as follows: increase the index by 1,
divide by the *new* index, and add a constant.

The integral of a sum (or difference) is also straightforward: if u and
v are functions of x then

$$\int (u+v)\,\mathrm{d}x = \int u\,\mathrm{d}x + \int v\,\mathrm{d}x$$

Finally, two other simple functions occurred as derivatives in Chapter
3 and consequently can readily be integrated:

$$\int \frac{\mathrm{d}x}{x} = \ln x + \alpha$$

$$\int \exp(x)\,\mathrm{d}x = \exp(x) + \alpha$$

Although rules exist for integrating products and other functions,
these are by no means as easy to apply as the rules for differentiating
the same sorts of function. The need for them occurs so infrequently
in elementary biochemistry that they are best dealt with when they do
by looking them up in standard tables (see below). There are, however,
a few functions that occur so often in kinetics that it is useful to be able
to recognize them. The first is the following:

$$\int \frac{\mathrm{d}x}{A+Bx} = \frac{1}{B}\ln(A+Bx) + \alpha$$

This can readily be confirmed by differentiating the right-hand side as a function of a function. Products of such functions also occur in kinetics and can be integrated by realizing that they can be expressed as sums of simple fractions. Thus:

$$\frac{1}{(a_1 + b_1 x)(a_2 + b_2 x)} \equiv \frac{A_1}{a_1 + b_1 x} + \frac{A_2}{a_2 + b_2 x}$$

where A_1 and A_2 are unknown constants such that

$$A_1(a_2 + b_2 x) + A_2(a_1 + b_1 x) \equiv 1$$

The *identity sign* ($\equiv$) emphasizes that the relationship must apply regardless of the value of x, and we can use this requirement to assign values to the two constants. For example, because the relationship must apply when $x = 0$ and also when $x = 1$ it follows that

$$A_1 a_2 + A_2 a_1 = 1$$

and

$$A_1(a_2 + b_2) + A_2(a_1 + b_1) = 1$$

Thus we have two simultaneous equations that can be solved for the unknowns A_1 and A_2. The two fractions produced in this way, which are known as *partial fractions*, can be integrated separately as above.

Another integral that is often needed in kinetic problems is the following:

$$\int \frac{x \, dx}{A + Bx}$$

The easiest way to deal with this is to define $u = A + Bx$, so that

$$du = B \, dx \quad \text{and} \quad x = (u - A)/B$$

and then

$$\int \frac{x \, dx}{A + Bx} = \frac{1}{B^2} \int \frac{(u - A) \, du}{u}$$

$$= \frac{1}{B^2} \left[\int du - A \int \frac{du}{u} \right]$$

$$= \frac{1}{B^2} \left[u - A \ln u \right] + \alpha$$

$$= \frac{1}{B^2}\left[A + Bx - A\ln(A + Bx)\right] + \alpha$$

Although this result is not in itself worth memorizing, the method by which it was obtained, in which we found a suitable function u that allowed a much easier integration, is often useful in integration problems.

The main results of this section are given in Table 4.1.

Table 4.1 Common integrals

$$\int Ax^i\,\mathrm{d}x = \frac{Ax^{i+1}}{i+1} + \alpha \qquad \text{(for any value of } i \textit{ except } -1)$$

$$\int (u + v)\,\mathrm{d}x = \int u\,\mathrm{d}x + \int v\,\mathrm{d}x + \alpha$$

$$\int \frac{\mathrm{d}x}{x} = \ln x + \alpha$$

$$\int \exp(Ax)\,\mathrm{d}x = \frac{1}{A}\exp(Ax) + \alpha$$

$$\int \frac{\mathrm{d}x}{A + Bx} = \frac{1}{B}\ln(A + Bx) + \alpha$$

$$\int \frac{x\,\mathrm{d}x}{A + Bx} = \frac{1}{B^2}\left[A + Bx - A\ln(A + Bx)\right] + \alpha$$

In all cases x is a variable, A, B and i are constants, u and v are functions of x, and α is a constant of integration. All of the integrals in this table apart from the last are worth memorizing.

4.4   Integral of $1/x$

Although I have asserted that $\dfrac{\mathrm{d}}{\mathrm{d}x}(\ln x) = \dfrac{1}{x}$ and, conversely, that $\displaystyle\int \frac{\mathrm{d}x}{x} = \ln x + \alpha$, I have not proved that this is so. A rigorous proof is hardly essential in an elementary course but it is useful to invoke plausibility arguments to increase one's sense of rightness for this important relationship. Fig. 4.2 represents the function $y = 1/x$. Let us now *define* the definite integral between 1 and some number a as

$$\int_1^a \frac{\mathrm{d}x}{x} = F(a)$$

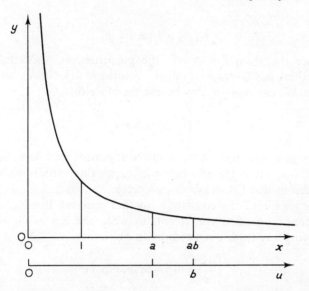

Fig. 4.2.   Plot of the hyperbola $y = 1/x$. The u-axis is an alternative abscissa axis related to the x-axis such that $u = x/a$. The area under the curve between $x = a$ and $x = ab$ is unaffected by calling it the area between $u = 1$ and $u = b$.

without specifying anything about the nature of $F(a)$. Then the area bounded by the curve, the x-axis and the vertical lines $x = 1$ and $x = a$ is given by

$$\int_1^a \frac{\mathrm{d}x}{x} = F(a)$$

and the area between $x = 1$ and $x = ab$ is given by

$$\int_1^{ab} \frac{\mathrm{d}x}{x} = F(ab)$$

which can equally well be written as

$$F(ab) = \int_1^a \frac{\mathrm{d}x}{x} + \int_a^{ab} \frac{\mathrm{d}x}{x} = F(a) + \int_a^{ab} \frac{\mathrm{d}x}{x}$$

Let us now define a new variable u such that $u = x/a$. Then $u = 1$ when $x = a$ and so

$$\int_a^{ab} \frac{\mathrm{d}x}{x} = \int_1^b \frac{\mathrm{d}u}{u} = F(b)$$

So
$$F(ab) = F(a) + F(b)$$

We see that $F(x)$ has exactly the properties we associate with logarithms: *addition* of $F(x)$ values is equivalent to *multiplication* of x values. We can more readily believe therefore that

$$\int \frac{dx}{x} = \ln x + \alpha$$

(although it does not follow from the argument used here that the result is $\ln x$ rather than, for example, $\log_{10} x$; this is true nonetheless, for reasons that I shall not discuss here).

The logarithm of a constant is another constant. It is legitimate, therefore, and often convenient, to write the last result in the following way (with α replaced by $\ln A$):

$$\int \frac{dx}{x} = \ln x + \ln A = \ln Ax$$

Accordingly one should not be surprised to see constants of integration appearing within logarithmic expressions as factors rather than as constants to be added to the expression as a whole.

4.5 Differential equations

One of the commonest reasons why one needs to integrate expressions in science is that scientific laws frequently make predictions about the *derivatives* of functions of interest rather than about the functions themselves. Consider, for example, a reaction between two molecules A and B:

$$A + B \rightarrow P$$

Elementary kinetic considerations suggest that if the mechanism is simple the rate of reaction is likely to be proportional to the concentration of A and also to the concentration of B, i.e.

$$\frac{dp}{dt} = kab$$

where a, b and p are the concentrations of A, B and P, respectively, at time t and k is a constant. Although this relationship is easy to arrive at and easy to understand it does not tell us very much as it stands; in particular, it does not tell us the extent of reaction as a function of time. To convert it into a more informative form we must first decrease the number of variables from 4 to 2. This can be done by means of the stoichiometric requirement that $(a + p)$ and $(b + p)$ must both be constants. If we define these as a_0 and b_0 respectively (the values of a and b when $p = 0$), the original equation can be written

$$\frac{dp}{dt} = k(a_0 - p)(b_0 - p)$$

This is an example of a simple *differential equation*. The easiest way of solving it is to *separate* the two variables so that only one appears on each side of the equation:

$$\int \frac{dp}{(a_0 - p)(b_0 - p)} = \int k\,dt$$

Solving the differential equation thus resolves itself into a problem of solving two integrals. The right-hand side is trivial and the left-hand side may be solved by partial fractions, so that

$$-\frac{\ln(a_0 - p)}{a_0 - b_0} + \frac{\ln(b_0 - p)}{a_0 - b_0} = kt + \alpha$$

Putting $p = 0$ when $t = 0$ we can evaluate the constant of integration α and rearrange the equation to

$$\ln\left[\frac{a_0(b_0 - p)}{b_0(a_0 - p)}\right] = (a_0 - b_0)kt$$

or

$$\frac{a_0(b_0 - p)}{b_0(a_0 - p)} = \exp[(a_0 - b_0)kt]$$

which is the solution of the original differential equation as it contains no derivatives.

Not all differential equations permit the variables to be separated in this way. Ones that do not are not common in elementary biochemistry, however, and so it is not worth while devoting much study to methods of solving them. It is much more efficient if one does not plan to be a specialist in a branch of biochemistry that requires differential equations to be solved frequently to seek expert help rather than labour at a problem that may not even be soluble at all. It is always wise to remember that even quite simple kinetic models can lead to intractable mathematics and there is little point in wasting time over an impossible problem. For example, the simplest model commonly discussed in enzyme catalysis is the Michaelis–Menten mechanism:

$$
\begin{array}{ccccccc}
 & & k_{+1} & & k_{+2} & & \\
E & + & S & \rightleftharpoons & ES & \rightleftharpoons & E & + & P \\
 & & k_{-1} & & k_{-2} & & \\
e_0 - x & & s_0 - x - p & & x & & & & p
\end{array}
$$

With concentrations defined as indicated under the equation application of elementary kinetic laws leads without difficulty to the following pair of simultaneous differential equations:

$$
\frac{dx}{dt} = k_{+1}(e_0 - x)(s_0 - x - p) - k_{-1}x - k_{+2}x + k_{-2}(e_0 - x)p
$$

$$
\frac{dp}{dt} = k_{+2}x - k_{-2}(e_0 - x)p
$$

Although it is not difficult to remove x from these equations by using the second one to express x and dx/dt in terms of p and its derivatives the resulting equation in p and t has no known solution.

If one arrives at an equation that is impossible to solve there is little point in relying on one's mathematical knowledge and ability. Instead, one has the choice between trying to solve it *numerically* rather than algebraically, an approach that is largely outside the scope of this book, and restricting the problem so as to make it soluble. This latter approach is very important and it demands scientific rather than mathematical knowledge. Thus, a mathematician might point out that the problem discussed above becomes soluble if $k_{+1} = k_{-2}$, but the biochemist would most likely regard this as an unreasonable restriction and would prefer to look for a restriction that could be made to be true by suitable choice of experimental conditions. By far the most commonly invoked restriction in enzyme kinetics is to require s_0 to be much larger than e_0 and to consider only the time scale in which dx/dt can be regarded as negligible. This is the basic restriction used in *steady-state kinetics*, but it is not the only possible one: alternatives might be to assume $s_0 \gg e_0$ and p negligible, the *pre-steady-state* condition, or to assume $e_0 \gg s_0$, the *single-turnover* condition. My purpose in mentioning these, however, is not to derive any kinetic results, but to emphasize that the decisions to be made are scientific, not mathematical: one needs some knowledge of enzymes and techniques for studying them to know that it is usually easy and convenient to achieve steady-state conditions but that assuming $k_{+1} = k_{-2}$ is not likely to be useful.

To summarize this section, the most efficient approach to differential equations in biochemistry is to proceed as follows: first, check whether the variables can be decreased to two that can be separated on to the two sides of the equation. If so, it can be solved in principle by integrating the two sides. If this cannot be done immediately then tables of integrals or expert advice are likely to prove much more rapid than blind struggling. If the variables cannot be separated or the problem appears impossible for some other reason, then try to introduce biochemically sensible restrictions or approximations that allow the simplified equation to be solved. If all else fails then seek expert help, remembering that a mathematician will probably know much less biochemistry than you do and will need some guidance over matters that you regard as obvious.

Finally, a word about tables of integrals and differential equations. Tables of integrals are easy to use, provided you remember two things: first, constants of integration are usually omitted, on the assumption that the user knows enough mathematics to supply a constant of integration without constant reminders; second, the symbol 'log' means 'ln' not, as in biochemical practice, '$\log_{10}$'. Tables of differential equations are much more difficult to use as they commonly assume some understanding of the principles of solving differential equations.

4.6 Numerical integration: evaluating the area under a curve

It sometimes happens that we need to know the area under a curve but cannot do the necessary integration exactly, either because we do not have an analytical expression for the curve or because we cannot integrate it. For example, the curve in Fig. 4.3 might represent the absorbance of the effluent from a chromatographic column as a function of the volume: if the absorbance of a sample is proportional to the concentration of some substance of interest, the total amount of the substance is proportional to the area under the curve. If it is not possible or convenient to integrate the curve analytically, the simplest

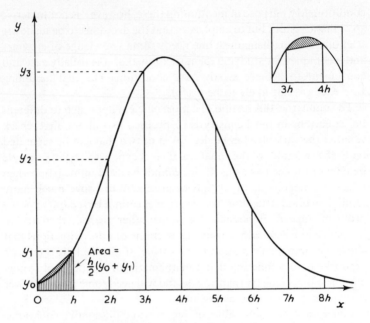

Fig. 4.3. Trapezium method for estimating the area under a curve. The shaded area between $x = 0$ and $x = h$ is $\frac{h}{2}(y_0 + y_1)$, but this is the area of the complete trapezium and thus significantly overestimates the area under the curve, which ought to exclude the diagonally shaded segment. Similar inaccuracies result from treating the other strips as trapezia, and the inset shows an example of significant underestimation. The alternative method discussed in the text, 'Simpson's rule', is much more accurate and the parabolic arcs that it assumes are hard to distinguish from the true curve, and for that reason are not shown in the figure.

way of estimating the area is to divide it into n parallel strips as shown and replace the short arcs between the ends of the strips by straight-line segments. It is then a matter of simple geometry to determine the areas of the trapezia produced. For example, between $x_0 = 0$ and $x_1 = h$, the area of the rectangle below the curve is $y_0 h$ and the area of the triangle between the rectangle and the curve is $(h/2)(y_1 - y_0)$, so the area of the whole trapezium is $(h/2)(y_0 + y_1)$. Similarly, the area of the adjacent trapezium is $(h/2)(y_1 + y_2)$, and so on. Every y except the first and last appears in the expressions for two adjacent trapezia, so the total area A under the straight-line segments is given by

$$A = \tfrac{1}{2}y_0 + y_1 + y_2 + y_3 + \ldots + y_{n-1} + \tfrac{1}{2}y_n$$

Although simple, this formula is not very accurate (unless n is made so large that the calculation becomes very tedious) because the straight-line segments leave appreciable areas unaccounted for, as illustrated in the inset to Fig. 4.3.

A more accurate method is to assume that over a limited range the curve can be approximated by a parabola. Thus instead of the unknown (or known but unintegrable) function $y = f(x)$ we write a quadratic

$$y \simeq a + bx + cx^2$$

If we evaluate y at three values of x, $x_0 = 0$, $x_1 = h$ and $x_2 = 2h$, and treat the approximation as exact, we have

$$y_0 = a$$
$$y_1 = a + bh + ch^2$$
$$y_2 = a + 2bh + 4ch^2$$

Eliminating a by subtraction, we have

$$y_1 - y_0 = bh + ch^2$$
$$y_2 - y_1 = bh + 3ch^2$$

and bh by further subtraction,

$$y_2 - 2y_1 + y_0 = 2ch^2$$

So

$$ch^2 = (y_2 - 2y_1 + y_0)/2$$

and

$$bh = y_1 - y_0 - ch^2 = 2y_1 - \tfrac{3}{2}y_0 - \tfrac{1}{2}y_2$$

Integrating the original quadratic, and substituting these values of a, bh and ch^2, we have

$$\int_0^{2h} y\,dx = \left[ax + \tfrac{1}{2}bx^2 + \tfrac{1}{3}cx^3 \right]_0^{2h} = 2ah + 2bh^2 + \tfrac{8}{3}ch^3$$
$$= \frac{h}{3}(y_0 + 4y_1 + y_2)$$

The area of the next pair of strips is similarly

$$\int_{2h}^{4h} y\,dx = \frac{h}{3}(y_2 + 4y_3 + y_4)$$

and so on. Notice that every odd-numbered y appears only once, whereas every even-numbered y apart from y_0 and y_n appears twice, so the total area A is given by

$$A = \frac{h}{3} \left[y_0 + 4(y_1 + y_3 + \ldots + y_{n-1}) + 2(y_2 + y_4 + \ldots + y_{n-2}) + y_n \right]$$

This formula is known as *Simpson's Rule*. It is hardly more difficult to apply than the trapezium method I considered first, and it is much more accurate. For most curves that are not highly complex the formula is adequately accurate if the number of strips is about ten, and in most experimental examples inaccuracies in measuring the y values is a more important source of error than the formula.

4.7 Problems

(4.1) Integrate the following:

(a) $\int (x^2 + 3x + 1) dx$

(b) $\int \left(x + \frac{1}{x} \right) dx$

(c) $\int \frac{dx}{2 + 3x}$

(d) $\int \exp(-3t) dt$

(4.2) Evaluate the following definite integrals:

(a) $\int_0^5 3x \, dx$

(b) $\int_1^2 \frac{dx}{x}$

(c) $\int_{-1}^1 3x^2 dx$

(d) $\int_{-2}^2 x^3 dx$

(4.3) Express the following equation in the form of partial fractions and use the result to evaluate $\int y \, dx$:

$$y(3x + 1)(x + 2) = 5$$

(4.4) Integrate the following expression:

$$\int \frac{(A + Bx)dx}{C + Dx}$$

(Note: this can most easily be done by separating it into two integrals.)

(4.5) The Michaelis–Menten equation can be written in the following form:

$$\frac{dp}{dt} = \frac{Vs}{K_m + s}$$

in which p and s are the concentrations of product and substrate, respectively, at time t, and V and K_m are constants. Defining $s = s_0$ and $p = 0$ at $t = 0$, use the one-to-one stoichiometry of the reaction to eliminate s and solve the resulting differential equation to get an expression for Vt in terms of p.

(4.6) The equation considered in problem (4.5) takes no account of the possibility of inhibition by the accumulating product. Show that the following equation, in which K_p is a constant and the other symbols are as in problem (4.5), can be integrated to provide an equation of the same form as the solution to problem (4.5):

$$\frac{dp}{dt} = \frac{Vs}{K_m(1 + p/K_p) + s}$$

What would be the form of a plot of a plot of $t/\ln[s_0/(s_0 - p)]$ against $p/\ln[s_0/(s_0 - p)]$?

(4.7) The relationship between the intensity of light I and the distance x through which it has passed in an absorbing solution can be expressed as follows:

$$\frac{dI}{dx} = -kcI$$

where k is a constant and c is the molar concentration of absorbing substance. Putting $I = I_0$ at $x = 0$, integrate this expression to obtain an expression for I in terms of x. If the *molar absorbance A* is defined such that

$$\log(I_0/I) = Acx$$

what is the relationship between A and k?

(4.8) In a spinning ultracentrifuge cell the outward flow rate of the mass of solute m is given, ignoring diffusion, by

$$\frac{dm}{dt} = \frac{\omega^2 x(1 - \overline{V}\rho)M_r A D c}{RT}$$

where t represents time, x the distance from the axis of rotation, M_r the relative molecular mass of the solute, A the cross-sectional area of the cell, D the diffusion coefficient, c the concentration of solute in g/l, and the other symbols represent constants that are either known or easily measurable. The effect of diffusion can be represented by

$$\frac{dm}{dt} = -AD\frac{dc}{dx}$$

After spinning for a long period the system achieves equilibrium and becomes time independent. Then the net flow rate, given by the sum of the two expressions, is zero at all points. Obtain an expression relating c and x at equilibrium and hence suggest a plot that would allow M_r to be conveniently measured.

(4.9) The following values of y at various values of x show measurements of the height of a recorder trace in a chromatographic separation. Estimate the areas under the two peaks by using Simpson's rule to evaluate $\int_1^9 y\,dx$ and $\int_9^{17} y\,dx$. In each case compare the results with four strips with those obtained with eight.

x	y	x	y	x	y
0	0.03	7	5.75	14	2.68
1	0.04	8	2.69	15	1.21
2	0.51	9	1.62	16	0.28
3	2.53	10	2.27	17	0.03
4	5.81	11	3.41	18	0.12
5	7.64	12	4.08	19	0.15
6	8.11	13	3.82	20	0.09

5 Solving Equations

5.1 Linear equations in one unknown

An equation such as

$$4 + x = 8 - 2x$$

contains only one unknown quantity, x, and is *linear*, because it contains no terms higher than the first power of x. Although it is easy to solve such an equation it is nonetheless useful to examine the processes involved as they illustrate in a simple way the methods to be used for solving more difficult equations. The first essential is to rearrange the equation so that terms of the same type, in this case constants and terms in x, are brought together. It is often convenient to collect all constants on the right-hand side and other terms on the left-hand side. This may be done by use of the principle that we can apply any operation we like to an equation (*other than* multiplying or dividing by zero) provided that we apply it identically to both sides. So, for example, we can subtract 4 from both sides of the above equation and add $2x$ to both sides:

$$4 + x - 4 + 2x = 8 - 2x - 4 + 2x$$

i.e.

$$3x = 4$$

The operation of subtracting, for example, 4 from both sides is clearly equivalent to 'bringing it from one side to the other of the equation and changing its sign'. Although this is a convenient and proper way of carrying out the operation one should realize that it is no more than an example of applying the same operation to both sides of an equation.

Continuing with the above example we may divide both sides of the last equation by 3 to obtain the solution:

$$x = 4/3 = 1.33 \text{ approx.}$$

Equations that do not seem to be linear at first sight can often be rearranged into linear form by application of the same methods. For example, the equation

$$\frac{8x}{5 + 2x} = 3$$

becomes linear after both sides of the equation are multiplied by $5 + 2x$:

$$8x = 15 + 6x$$

which can be solved as before to give $x = 7.5$.

The only difficulty that may arise in solving this sort of equation is that it may turn out to be *singular*, which means that it contains less information than it seems to contain. Consider, for example, the following equation, which appears to be an ordinary linear equation:

$$5x + 2 - 3(x + 8) = 2x - 22$$

If, however, we collect terms, we find

$$(5 - 3 - 2)x = -22 - 2 + 24$$

i.e. $0x = 0$, which is true regardless of the value of x and consequently tells us nothing about the value of x. Examples as simple as this seem rather artificial and trivial, but singular equations can be a serious problem because they can appear in the last stages of solving a more complex equation or set of equations. This often happens in the analysis of experiments that have been badly planned so that they do not in fact provide the information they were intended to provide, and I shall consider some more realistic examples later in this chapter.

5.2 Rearranging equations

Equations in science often contain the information we require but in inverse form, i.e. instead of giving the required unknown in terms of known quantities an equation may express a known quantity in terms of the unknown. In such cases we can often arrive at a more useful formulation of the equation by solving it in the same way as we would if it were expressed numerically.

For example, in studies of individual enzymes it is usual to regard the rate of reaction v as something that is determined by the properties of the enzyme and by the concentrations of relevant species such as the enzyme and its substrate. We might write, for example,

$$v = \frac{k_{cat}e_0 s}{K_m + s}$$

where e_0 and s are the concentrations of enzyme and substrate, respectively, and k_{cat} and K_m are constants. When we consider the enzyme not as an isolated catalyst but as a component of a metabolic pathway, however, we might argue that it is not s that determines v, but v that determines s, because v is decided by the flux through the whole pathway whereas s reflects the capacity of the enzyme to remove the particular substrate. If the enzyme is very active, s will be small; if it is not very active, s may be large; but provided it is not saturated with substrate the value of v will be decided by the rate at which other enzymes are supplying the substrate. Accordingly, we may ask, how can s be expressed in terms of v? This is exactly analogous to the second example in the previous section except that now we must deal with constants expressed as symbols rather than as numbers. The method is the same: we multiply by $K_m + s$,

$$v(K_m + s) = k_{cat}e_0 s$$

collect terms in s on the left-hand side and other terms on the right-hand side,

$$vs - k_{cat}e_0 s = -vK_m$$

and divide by the coefficient of s, $(v - k_{cat}e_0)$, to give the solution:

$$s = \frac{K_m v}{k_{cat}e_0 - v}$$

5.3 Simultaneous linear equations

If we have two or more unknown quantities a single equation is insufficient for defining their values. For example, the following equation:

$$2x + 3y = 12$$

can be satisfied by $x = 3$, $y = 2$, but this is not a unique solution and there are an infinity of other possible solutions, such as $x = 4.5$, $y = 1$. In fact regardless of the value of x there is a value of y given by $(12 - 2x)/3$ such that x and y satisfy the equation exactly. When there are fewer equations than unknowns the problem is said to be *underdetermined*. For there to be a unique solution the number of equations must be equal to the number of unknowns. The reason for this is that each equation can be used to express one unknown in terms of the others but once that has been done no further information is available from that equation. If one loses an equation for each unknown that is eliminated it is clear that one can only finish with a single equation in a single unknown if the numbers of equations and unknowns were initially the same. Although the approach is general I shall consider only the case of two equations in two unknowns, because it is rare in elementary biochemistry for one to have problems with three or more unknowns. Such equations are called *simultaneous equations* because they define two (or more) relationships that are true simultaneously.

Let us apply the approach indicated above to the following pair of equations:

$$2x + 3y = 12$$

$$2x + y = 8$$

Treating the first equation as a linear equation in y with x as an unknown constant, we have

$$3y = 12 - 2x$$

Hence

$$y = 4 - 2x/3$$

Substituting this for y in the second equation (the first being of no further use to us) we have

$$2x + 4 - 2x/3 = 8$$

which may be solved as a simple linear equation in x to yield $x = 3$, and, after substitution back into the expression for y, $y = 2$. This approach is superficially attractive because it can be applied mechanically, i.e. without thought, and it ought always to work. Nonetheless it is not the best way of solving simultaneous equations and one can often proceed more simply and with less likelihood of error by thinking first.

For the particular example we have considered at least two improvements to the above method are possible. First, we ought to notice immediately by inspection that the second equation gives a simpler expression for y than the first does, i.e.

$$y = 8 - 2x$$

and by substituting this into the first equation we can avoid fractions:

$$2x + 3(8 - 2x) = 12$$

etc., with the same solution.

Second, we should also notice from inspection that x is multiplied by the same constant 2 in both equations. Consequently, we can eliminate x immediately by subtracting the second equation from the first:

$$
\begin{array}{r}
2x + 3y = 12 \\
2x + y = 8 \\
\hline
2y = 4
\end{array}
$$

Hence $y = 2$, etc. The subtraction of one equation from another is justified by the fact that *if* $2x + y = 8$ then subtracting $2x + y$ is equivalent to subtracting 8 and so by subtracting $2x + y$ from the left-hand side and 8 from the right-hand side we are applying identical operations to the two sides of the equation.

This third approach is clearly the simplest and quickest of the three, but it seems to require the coincidence of finding the same term $-2x$ in this example – in both equations. In fact we can always ensure that such coincidences occur by multiplying by appropriate factors in the first instance. For example, if

$$
\begin{array}{r}
3x - 5y = 3 \\
2x + 3y = 21
\end{array}
$$

then multiplying the first equation by 2, the coefficient of x in the second equation, and the second equation by 3, the coefficient of x in the first equation, ensures that x has the same coefficient in both equations:

$$
\begin{array}{r}
6x - 10y = 6 \\
6x + 9y = 63
\end{array}
$$

from which we can proceed as before to obtain the solution, $x = 6$, $y = 3$.

An alternative way of achieving the same effect is to *divide* each equation by the coefficient of x in *that* equation (or whatever unknown is chosen for elimination). In the above example, we would have:

$$x - 1.667y = 1$$
$$x + 1.500y = 10.5$$

As this approach brings about the appearance of non-integral coefficients it may seem rather unappealing, but it is nonetheless a good general approach. One reason for this is that in experimental problems (as opposed to invented ones in textbooks) the coefficients are hardly ever integral in the first place. Another is that it tends to ensure that the calculation is done in terms of numbers not very different from 1. This tends to decrease numerical inaccuracies, whether the calculation is done by hand or by computer. Although these numerical considerations may seem trivial, and indeed usually are trivial for simultaneous equations in only two unknowns, they become of great importance in solving sets of equations in several unknowns.

5.4   Determinants

The methods I have described for solving simultaneous equations expressed in numerical terms can also be applied to equations in which the coefficients are not assigned numerical values. Consider the following general pair of linear simultaneous equations:

$$A_1 x + B_1 y + C_1 = 0$$
$$A_2 x + B_2 y + C_2 = 0$$

Multiplying the first by A_2 and the second by A_1 we have

$$A_1 A_2 x + A_2 B_1 y + A_2 C_1 = 0$$
$$A_1 A_2 x + A_1 B_2 y + A_1 C_2 = 0$$

and after subtraction of the first equation from the second,

$$(A_1 B_2 - A_2 B_1)y + (A_1 C_2 - A_2 C_1) = 0$$

Hence

$$-y = \frac{A_1 C_2 - A_2 C_1}{A_1 B_2 - A_2 B_1}$$

and, similarly,

$$x = \frac{B_1 C_2 - B_2 C_1}{A_1 B_2 - A_2 B_1}$$

A special notation has been devised to allow this last result to be expressed in a particularly elegant way, as follows:

$$\frac{x}{\begin{vmatrix} B_1 & C_1 \\ B_2 & C_2 \end{vmatrix}} = \frac{-y}{\begin{vmatrix} A_1 & C_1 \\ A_2 & C_2 \end{vmatrix}} = \frac{1}{\begin{vmatrix} A_1 & B_1 \\ A_2 & B_2 \end{vmatrix}}$$

The denominators of these expressions are called *determinants*. Any determinant is written as a square array of numbers or algebraic symbols between vertical lines and it has a numerical or algebraic value that can be found by applying a general rule. Although one can have any size of square I shall only consider 2 × 2 determinants, which are not only particularly simple but are also the most important for our purposes: the value of such a determinant is the product of the top-left and bottom-right elements minus the product of the bottom-left

$$\begin{vmatrix} a_1 & b_1 \\ a_2 & b_2 \end{vmatrix} \equiv \begin{pmatrix} a_1 & \\ & b_2 \end{pmatrix} - \begin{pmatrix} & b_1 \\ a_2 & \end{pmatrix} = a_1 b_2 - a_2 b_1$$

To write the solution of a set of equations in determinant form we proceed in accordance with the following set of rules. First, the original set of equations must be organized systematically, so that the unknowns appear in the same sequence in every equation (with appropriate gaps or zeroes if there are any unknowns missing from particular equations) and the constant as the last term on the *left-hand* side of each equation. For each equation the right-hand side should then consist of zero. Each expression except the last in the set of equations representing the solution consists of a fraction with a different unknown in each numerator and a determinant in each denominator. For each unknown the determinant in the denominator consists of all the coefficients in the original equation, in the same order, *except* for those of the unknown in the numerator. The last (right-hand) expression is also a fraction, with 1 in the numerator and a denominator consisting of a determinant containing all of the coefficients in the original equations but not the constants. The left-hand expression has a positive sign and the remainder alternate in sign on reading from left to right. Applying these rules to the following pair of equations:

$$2x + 3y = 5$$
$$y - 5 = 4x$$

we first rearrange the equations into standard form,

$$2x + 3y - 5 = 0$$
$$4x - y + 5 = 0$$

and then write down the solution in determinant form:

$$\frac{x}{\begin{vmatrix} 3 & -5 \\ -1 & 5 \end{vmatrix}} = \frac{-y}{\begin{vmatrix} 2 & -5 \\ 4 & 5 \end{vmatrix}} = \frac{1}{\begin{vmatrix} 2 & 3 \\ 4 & -1 \end{vmatrix}}$$

or

$$x/10 = -y/30 = -1/14$$

so $x = -10/14 = -5/7$, $y = 30/14 = 15/7$.

It often seems more natural to write each of the original set of equations with the constant on the right-hand side rather than as the last term on the left-hand side, i.e. (for the same example):

$$2x + 3y = 5$$
$$4x - y = -5$$

If this is done the solution can be written in the same way as before *except* that the sign of the constant is the opposite of what it would be if the equations were written in standard form. For two simultaneous equations this means that the last numerator is -1 rather than 1:

$$\frac{x}{\begin{vmatrix} 3 & 5 \\ -1 & -5 \end{vmatrix}} = \frac{-y}{\begin{vmatrix} 2 & 5 \\ 4 & -5 \end{vmatrix}} = \frac{-1}{\begin{vmatrix} 2 & 3 \\ 4 & -1 \end{vmatrix}}$$

The determinant in the right-hand (constant) expression is especially important for any set of equations because it appears in the solution for every unknown. Moreover, its value indicates whether a unique solution to the set of equations exists: if it is zero there is no unique solution and the equations are said to be *singular*. It is therefore useful to be able to recognize characteristics of a determinant whose value is zero (these apply to all determinants, not just those of order 2). First, if any row or column consists entirely of zeroes the determinant is zero. This is always true for a set of equations in which the number of equations is different from the number of unknowns. Second, if any two rows or any two columns are identical the determinant is zero. This occurs if any equation is identical to any other, so that the number of *different* equations is less than the number of unknowns, or if the information given about any two unknowns is identical, so that they cannot be distinguished. Third, if any row or column can be obtained by multiplying all of the elements in another row or column by the same constant the determinant is zero. This corresponds to the case where one equation can be obtained from another merely by multiplying every term by the same constant. All of these characteristics that cause determinants to be zero can be recognized immediately by inspection, but there is a fourth that is less obvious. It is not possible at all with only two simultaneous equations but can render larger sets of equations insoluble: if any row or column is a linear function of two or more other rows or columns the determinant is zero. For example, the determinant

$$\begin{vmatrix} 2 & 3 & 7 \\ 1 & 4 & 6 \\ 0 & 1 & 1 \end{vmatrix}$$

can be seen to have a value of zero by the fact that each element in the third column is equal to twice the corresponding element in the first column plus the corresponding element in the second column.

5.5 Quadratic equations

Suppose we have the equation

$$(x - 5)(x - 2) = 0$$

then it is obvious that the left-hand side is equal to zero if *either* of the two terms in parenthesis is zero, i.e. $x - 5 = 0$ *or* $x - 2 = 0$, because anything multiplied by zero is zero. Thus there are two values of x, $x = 5$ and $x = 2$, that satisfy the equation. If we multiply out the left-hand side we have a typical *quadratic equation*:

$$x^2 - 7x + 10 = 0$$

which is identical to the original equation and hence must have the same pair of solutions, $x = 5$ or $x = 2$.

In principle, a quadratic equation can be solved by reversing the above process, i.e. by finding a pair of factors that give the left-hand side when multiplied together. In practice most of the equations we meet in science cannot be solved in this way because no *rational factors* exist. (Strictly, a rational number is one that can be expressed as a ratio of two integers, but it will do little harm if we think of it loosely as a 'simple' number.) Nonetheless, we can apply the same sort of method to obtain a general solution to the problem of how to solve a quadratic equation. Consider the following equation:

$$ax^2 + bx + c = 0$$

which we can write as

$$x^2 + (b/a)x + (c/a) = 0$$

Let us compare the left-hand side of this with the square of $(x + b/2a)$, which is

$$(x + b/2a)^2 = x^2 + (b/a)x + b^2/4a^2$$

Apart from the third term this is the same as the left-hand side of the equation to be solved, so, subtracting the latter (which has a value of zero) we have

$$(x + b/2a)^2 = b^2/4a^2 - c/a = (b^2 - 4ac)/4a^2$$

and hence

$$x + b/2a = \frac{\pm \sqrt{(b^2 - 4ac)}}{2a}$$

which rearranges to

$$x = \frac{-b \pm \sqrt{(b^2 - 4ac)}}{2a}$$

We use the sign $\pm$ because *either* of the two expressions with square equal to $(b^2 - 4ac)$ will satisfy the equation: remember that the square-root sign is not ambiguous and specifies the positive square root *only*. This result is the general solution for a quadratic equation. Whether it is worth memorizing depends on how often you plan to solve quadratic equations, but it is important to understand how to use it even if you do not memorize it.

The quantity $(b^2 - 4ac)$ is known as the *discriminant* of the equation $ax^2 + bx + c = 0$. It is a useful quantity because its value indicates the nature of the solutions of the equation. If it is positive, the equation has two real and unequal solutions; if it is zero there are two real solutions, but they are identical; if it is positive and a perfect square, the solutions are rational and the simple factorization method mentioned at the beginning of this section could be used for solving the equation; if it is negative there are no real solutions. This last result follows from the fact that a negative number does not have a real square root, and it is important because it shows that one can have an apparently elementary equation that has no solution. For elementary purposes we can omit the word 'real' from the preceding statements, because in elementary work only real solutions need be considered as solutions at all. In more advanced mathematics it becomes convenient to define so-called 'imaginary' numbers as the square roots of negative real numbers, and 'complex' numbers as the results of adding real and imaginary numbers together. However, although these kinds of numbers have important applications in some scientific contexts – in quantum mechanics and in the theory of electrical circuits, for example – it is difficult to think of an application in elementary biochemistry and so I shall not consider them further in this book.

It often happens when quadratic equations occur in scientific contexts that although there are two solutions mathematically only one of them is physically meaningful, and this can usually be identified by inspection. Consider, for example, the equilibrium between the three adenylates:

$$2\,ADP \rightleftharpoons ATP + AMP$$

which has an equilibrium constant of about 1/2 under physiological conditions, i.e.

$$\frac{[ATP][AMP]}{[ADP]^2} = \frac{1}{2}$$

Suppose now we mix 8mM-ATP with 2mM-ADP in the presence of the enzyme adenylate kinase, which catalyses the equilibration: what will be the concentrations of the three adenylates at equilibrium? If we put $[AMP] = x$, the stoichiometry of the reaction requires that $[ATP] = 8 + x$, $[ADP] = 2 - 2x$, and so the equilibrium expression can be written as

$$\frac{(8 + x)x}{(2 - 2x)^2} = \frac{1}{2}$$

or, after rearranging,

$$x^2 - 12x + 2 = 0$$

The formula we derived above yields the following two solutions: $x = 11.8$ *or* 0.169mM. At first sight these are both physically meaningful if we simply consider them as AMP concentrations, but calculation of the corresponding ADP and ATP concentrations shows that

if $[AMP] = 11.8$mM, then $[ADP] = -21.6$mM, $[ATP] = 19.8$mM

if $[AMP] = 0.169$mM, then $[ADP] = 1.66$mM, $[ATP] = 8.17$mM

As concentrations in the real world cannot be negative, only the second solution is physically acceptable and the first must be discarded. As a general rule one should always check that the solution of an equation makes physical as well as mathematical sense. It is perfectly possible to have results that are entirely sound mathematically but are nonetheless physically nonsensical.

5.6 Graphical solution of equations

Sometimes we have to solve equations that are more complex than quadratics. Exact analytical solutions for certain kinds of higher-order equations exist, but these are little used and the practical choice is between *graphical solution* and *Newton's method*, which I shall discuss in the next two sections.

Let us consider the following equation as an example of the sort of equation that would be very difficult (if not impossible) to solve analytically:

$$x^2 + \ln x = 7$$

We can solve this quite easily by a graphical method if we define a function $f(x)$ as

$$f(x) = x^2 + \ln x - 7$$

This function must be zero when x is equal to the solution of the equation. If we plot $f(x)$ against x (Fig. 5.1), the line crosses the x-axis at about $x = 2.47$. As the x-axis is an expression of the relationship $f(x) = 0$ it follows that $x = 2.47$ is an approximate solution to the equation.

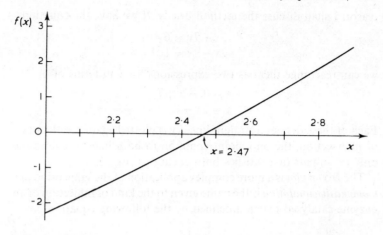

Fig. 5.1. Graphical solution of the equation $x^2 + \ln x = 7$. The line is a plot of the function $f(x) = x^2 + \ln x - 7$, which must be zero when the equation is satisfied. Thus $x = 2.47$ is an approximate solution.

The advantage of graphical solution is that it is easy to understand and use, but it has compensating disadvantages also. First, it is laborious, as one has to evaluate the function at numerous different x values, many of which turn out not to be close to the solution. Second, it is limited in accuracy by one's draughtsmanship and ability to read off the solution accurately. In principle, one could improve the accuracy of the above solution by drawing another graph on a larger scale covering only a narrow range of x values, say from 2.46 to 2.48. This would add substantially to the labour required, however, and one would normally do better to proceed to Newton's method (next section) if one needed a more accurate solution than the one given by the preliminary graph.

One can also use graphical methods for solving simultaneous equations. Although this is not a particularly convenient way of dealing with straightforward linear simultaneous equations of the sort I discussed earlier in this chapter, it is nonetheless important for two reasons. First, it provides one of the simplest ways of solving non-linear equations. Second, the method illustrates the principle underlying certain graphical techniques used in enzyme kinetics, most notably the *Dixon plot* for determining inhibition constants. For this reason I shall discuss the method briefly. If we have the equations

$$5x + 7y = 6$$
$$2x - y = 1$$

we can rearrange them as two expressions for y in terms of x:

$$y = (6 - 5x)/7$$
$$y = 2x - 1$$

Each of these defines a straight line if y is plotted against x; the point of intersection, the only point common to both lines, provides the only (x, y) pair that satisfies both equations (Fig. 5.2).

The *Dixon plot* is a more complex application of the same principle. *Competitive inhibition* is the name given to the kind of inhibition of an enzyme-catalysed reaction defined by the following equation:

$$v = \frac{Vs}{K_m(1 + i/K_i) + s}$$

in which v is the rate observed at concentrations i and s of inhibitor and substrate, respectively, and V, K_m, K_i are constants. The Dixon

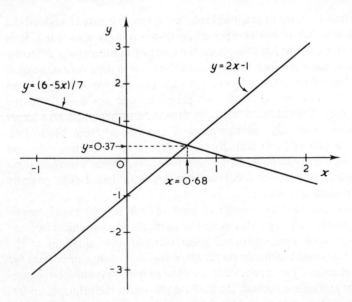

Fig. 5.2. Graphical solution of the simultaneous equations $5x + 7y = 6$, $2x - y = 1$. The co-ordinates of the point of intersection provide the only (x, y) pair that satisfies both equations, i.e. $x = 0.684$, $y = 0.368$.

plot is especially convenient in the common experimental circumstance where our main objective is to measure K_i, the other two parameters V and K_m being either of less immediate concern or known independently. If we have values of v measured at several i values at each of two s values, s_1 and s_2, we can determine K_i by plotting $1/v$ against i at both s values. If we define v_1 for the v values measured at $s = s_1$, the two straight lines are given by

$$1/v_1 = (K_m/Vs_1) + (1/V) + (K_m/VK_is_1)i$$
$$1/v_2 = (K_m/Vs_2) + (1/V) + (K_m/VK_is_2)i$$

The point of intersection of the two lines defines the value of i at which $1/v_1 = 1/v_2$ and hence the value at which the two right-hand sides are equal:

$$(K_m/Vs_1) + (1/V) + (K_m/VK_is_1)i = (K_m/Vs_2) + (1/V) + (K_m/VK_is_2)i$$

Despite its messy appearance this is just a simple linear equation in i which may be solved in the ordinary way to yield $i = -K_i$. Thus $-K_i$ is given by the value of i at the point of intersection of the two lines (Fig. 5.3).

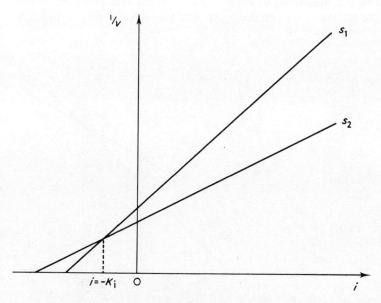

Fig. 5.3.   Dixon plot for determining a competitive inhibition constant: this is an application of the method illustrated in Fig. 5.2.

Another and somewhat simpler application of the graphical solution of simultaneous equations in enzyme kinetics is found in the *direct linear plot* for determining the parameters K_m and V in the Michaelis–Menten equation. If the rates v_1 and v_2 at substrate concentrations s_1 and s_2, respectively, are given by the following pair of equations:

$$v_1 = \frac{Vs_1}{K_m + s_1}$$

$$v_2 = \frac{Vs_2}{K_m + s_2}$$

then K_m and V can be evaluated graphically by rearranging the equations to show the dependence of V on K_m:

$$V = v_1 + (v_1/s_1)K_m$$

$$V = v_2 + (v_2/s_2)K_m$$

Both of these define straight lines when V is plotted against K_m. The first may most easily be drawn as a straight line with intercepts $-s_1$ on the K_m axis and v_1 on the V axis, and the second similarly (Fig. 5.4).

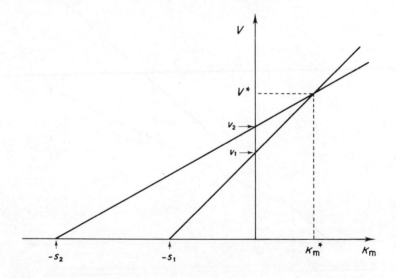

Fig. 5.4. Determination of the parameters of the Michaelis–Menten equation by the graphical method illustrated in Fig. 5.2.

The co-ordinates of the point of intersection give K_m and V. This plot is exactly analogous to the solution of simultaneous equations illustrated in Fig. 5.2.

5.7 Newton's method

Although graphical solution is easy to understand and carry out its accuracy is limited and it can be excessively laborious. For computational purposes *Newton's method* is much better. It depends on the fact that most smooth functions behave over short ranges in simple ways that can be predicted approximately from knowledge of the function value and its first derivative at particular points.

I shall use as an example the same equation as we used for studying the graphical method:

$$x^2 + \ln x = 7$$

Then if we define a function $f(x)$ as

$$f(x) = x^2 + \ln x - 7$$

we have, as before, solved the equation if we find a value of x for which $f(x) = 0$. At any arbitrary starting point x_0 the *tangent* to the graph of $f(x)$ against x is given by

$$y = f(x_0) + (x - x_0)f'(x_0)$$

where $f'(x_0)$ is the first derivative of $f(x)$ evaluated at $x = x_0$. As seen in Fig. 5.5, this tangent is approximately coincident with the curve representing the true function over a finite range of x values. Consequently, if we put $y = 0$ and solve for x we should get a value x_1 that is nearer the solution than x_0, i.e. we put

$$f(x_0) + (x_1 - x_0)f'(x_0) = 0$$

and rearrange to give

$$x_1 = x_0 - \frac{f(x_0)}{f'(x_0)}$$

This is the fundamental relationship of Newton's method. Let us see how it can be used for solving the equation that we set out with. For that equation, differentiation of $f(x)$ with respect to x shows that $f'(x)$ is given by

$$f'(x) = 2x + 1/x$$

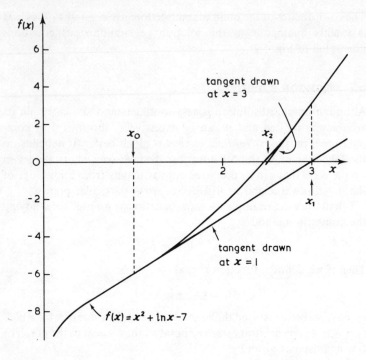

Fig. 5.5.   Newton's method for solving the equation $x^2 + \ln x - 7$ (see Fig. 5.1). The starting guess $x = x_0 = 1$ provides the tangent that crosses the x-axis at $x = x_1 = 3$. This improved guess for the solution provides a second tangent that crosses the axis at $x = x_2 = 2.51$. Further tangents calculated in the same way give, successively, $x = 2.469, 2.469\,04, 2.469\,042\,3\ldots$. This method permits much greater accuracy than one can hope to achieve with a graphical method.

If we choose an arbitrary starting point of $x_0 = 1$, then

$$x_1 = 1 + 6/3 = 3$$

We can see from Fig. 5.5 that although this is not the correct solution it is closer to it than $x_0 = 1$. So we proceed with a second approximation, replacing x_0 with x_1 and x_1 with x_2, and so on:

$$x_2 = 3 - 3.099/6.333 = 2.51$$
$$x_3 = 2.51 - 0.220/5.418 = 2.469$$
$$x_4 = 2.469 + 0.0002/5.3430 = 2.46904$$

One can continue until any desired accuracy is achieved.

In the above example one could recognize that the successive approximations were nearer and nearer to the correct solution by the fact that the numerical values (i.e. the values ignoring the signs) of $f(x)$ in the numerator of the second term were getting smaller and smaller with each approximation: $-6, 3.099, 0.220, -0.0002$. Sometimes this does not happen and instead of getting better and better the successive approximations given by Newton's method get worse and worse. This may happen, for example, if $f'(x_0)$ is close to zero, so that x_0 is close to a maximum or a minimum of $f(x)$. One can usually cure this problem when it occurs by trying a different value of x_0. In severe cases a preliminary graphical exploration as described in the preceding section should reveal a suitable range of values to try. For equations with two or more solutions, Newton's method does not guarantee to find all of them; it will usually proceed towards the solution closest to x_0. Again, graphical exploration is useful in such cases to ensure that no solutions have been overlooked.

5.8 Approximate methods

In textbook illustrations of methods for solving equations one commonly deals with examples in which the coefficients are numbers – often integers – in the range 1 to 10. Real life is often very different and one may instead have to solve equations with coefficients ranging over several orders of magnitude. This has two consequences: first, methods that work well with simple examples may work badly with real ones; and second, there is much more scope for the use of approximations than one might expect from the study of elementary examples.

To illustrate these points, let us consider how to calculate the pH of a 0.1M solution of monosodium glutamate, given that glutamic acid has three ionizations, with pK_a values of 2.3, 4.3 and 9.7. These convert to K_a values of 5×10^{-3}, 5×10^{-5} and 2×10^{-10}, respectively, and so the equilibria can be formulated as follows:

$$G^{1+} \xrightleftharpoons{5 \times 10^{-3}} G^0 \xrightleftharpoons{5 \times 10^{-5}} G^{1-} \xrightleftharpoons{2 \times 10^{-10}} G^{2-}$$

where G^{1+} represents protonated glutamic acid with a net charge of $+1$, G^0 represents glutamic acid with a net charge of zero, etc. (actually three different states with charge zero can be drawn, but for calculating pH values it is not necessary to distinguish between them, and G^0 comprises all of them; similarly, G^{1-} comprises three states with charge -1). From the definitions of the equilibrium constants we can write

$$[G^0] = 5 \times 10^{-3}[G^{1+}]/[H^+]$$

$$[G^{1-}] = 5 \times 10^{-5}[G^0]/[H^+] = 2.5 \times 10^{-7}[G^{1+}]/[H^+]^2$$

$$[G^{2-}] = 2 \times 10^{-10}[G^{1-}]/[H^+] = 5 \times 10^{-17}[G^{1+}]/[H^+]^3$$

From the stoichiometry we have the following conservation equation:

$$[G^{1+}] + [G^0] + [G^{1-}] + [G^{2-}] = 0.1M$$

and in addition, from the fact that the total concentration of negative charge must equal the concentration of Na^+ ions,

$$-[G^{1+}] + [G^{1-}] + 2[G^{2-}] = 0.1M$$

Substitution of the first three of these five expressions into the last two gives

$$[G^{1+}]\left(1 + \frac{5 \times 10^{-3}}{[H^+]} + \frac{2.5 \times 10^{-7}}{[H^+]^2} + \frac{5 \times 10^{-17}}{[H^+]^3}\right) = 0.1M$$

$$[G^{1+}]\left(-1 + \frac{2.5 \times 10^{-7}}{[H^+]^2} + \frac{1 \times 10^{-16}}{[H^+]^3}\right) = 0.1M$$

After eliminating $[G^{1+}]$ by dividing one equation by the other, multiplying through by $[H^+]^3$ and rearranging, we have the following cubic equation in $[H^+]$:

$$2[H^+]^3 + 5 \times 10^{-3}[H^+]^2 - 5 \times 10^{-17} = 0$$

On the face of it this is a straightforward cubic equation and we might expect to be able to solve it by Newton's method. However, the enormous range of coefficients from 2 to 5×10^{-17} makes it a very different proposition from the equations we have considered earlier in this chapter. If we define

$$f([H^+]) = 2[H^+]^3 + 5 \times 10^{-3}[H^+]^2 - 5 \times 10^{-17}$$

$$f'([H^+]) = 6[H^+]^2 + 10^{-2}[H^+]$$

and choose $[H^+]_0 = 1M$ as a (very poor) starting guess, we have

$$[H^+]_1 = 6.66 \times 10^{-1}\,M$$

$$[H^+]_2 = 4.44 \times 10^{-1} \, M$$

$$[H^+]_3 = 2.96 \times 10^{-1} \, M$$

.

.

.

.

Progress is in the right direction but is painfully slow. In fact it is not until the 30th approximation that we have even one significant figure correct:

$$[H^+]_{30} = 1.18 \times 10^{-7} \, M$$

$$[H^+]_{31} = 1.01 \times 10^{-7} \, M$$

$$[H^+]_{32} = 1.00 \times 10^{-7} \, M$$

All subsequent approximations are very close to this last, showing that the correct pH is 7.0.

We can see by inspection that the answer is equal to the mean of the two higher pK_a values, i.e. $7.0 = (4.3 + 9.7)/2$, and we might well enquire whether this is a coincidence, or whether we could have arrived at it by a quicker and easier route. We can indeed: the same enormous range of coefficients in the original equation that caused Newton's method to perform so badly allows us to introduce approximations that are virtually exact in this sort of problem even though the corresponding ones would be quite improper in an elementary example. To do this it is safest not to regard it as a purely mathematical problem but to make use of our knowledge of chemistry. Because of the 0.1 M concentration of Na^+ ions, we know that the average charge on the glutamic acid species must be -1, so it is reasonable to guess that the predominant species will be G^{1-} and that G^{1+} will be present at negligible concentration. If we ignore G^{1+}, a derivation along exactly the same lines as above leads to the equation

$$[H^+]^2 - 10^{-14} = 0$$

which yields the solution $[H^+] = 10^{-7} \, M$, or pH $= 7.00$, immediately.

The contrast between these two ways of solving the same problem could hardly be more striking. By applying a simple approximation at the beginning we were able to obtain a solution that is so nearly exact that it is scarcely an approximation at all. Virtually all pH problems can be dealt with in this way with little difficulty. Even if several ionizations are involved there are usually no more than three ionic states that need to be considered, and these can be identified by inspection. I shall illustrate this by reference to a more complex example, the pH of a solution of glutathione partially titrated with NaOH. Glutathione is a peptide (γ-glutamylcysteinylglycine) with four ionizable groups and hence four pK_a values, 2.12, 3.53, 8.66 and 9.12 (corresponding to K_a values of 7.59×10^{-3}, 2.95×10^{-4}, 2.19×10^{-9} and 7.59×10^{-10}, respectively). Suppose we add 2.5 ml of 1M NaOH to 10ml of 0.1M glutathione: what is the pH of the resulting mixture? The total glutathione concentration is $10 \times 0.1/12.5 = 0.08$M, so

$$[G^{1+}] + [G^0] + [G^{1-}] + [G^{2-}] + [G^{3-}] = 0.08M$$

where the various ionic states are represented by the same sort of symbolism as used above for the ionic states of glutamic acid; and the Na^+ concentration is $2.5/12.5 = 0.2$M, so

$$-[G^{1+}] + [G^{1-}] + 2[G^{2-}] + 3[G^{3-}] = 0.2M$$

If we continued by means of Newton's method we should obtain a quartic equation and solving it would be even more laborious than the cubic equation we had before. Instead, we shall proceed as follows: in pure glutathione the predominant form must be G^0 in order to achieve an average charge of zero; but 0.2M is 2.5 times 0.08M, i.e. the NaOH added was sufficient to neutralize more than two groups on G^0. We should expect, therefore, that the predominant forms in the mixture should be G^{2-} and G^{3-}. So let us ignore G^{1+} and G^0 altogether and write the above pair of equations as

$$[G^{1-}] + [G^{2-}] + [G^{3-}] = 0.08M$$

$$[G^{1-}] + 2[G^{2-}] + 3[G^{3-}] = 0.2M$$

Moreover,

$$[G^{2-}] = [G^{3-}][H^+]/7.59 \times 10^{-10}$$

$$[G^{1-}] = [G^{2-}][H^+]/2.19 \times 10^{-9} = [G^{3-}][H^+]^2/1.66 \times 10^{-18}$$

Hence

$$[G^{3-}]\left(1 + \frac{[H^+]}{7.59 \times 10^{-10}} + \frac{[H^+]^2}{1.66 \times 10^{-18}}\right) = 0.08M$$

$$[G^{3-}]\left(3 + \frac{2[H^+]}{7.59 \times 10^{-10}} + \frac{3[H^+]^2}{1.66 \times 10^{-18}}\right) = 0.2M$$

Dividing one expression by the other we can eliminate $[G^{3-}]$ and rearrange to the following quadratic equation for $[H^+]$:

$$7.23 \times 10^{16}[H^+]^2 + 5.30 \times 10^7[H^+] - 0.04 = 0$$

which may be solved like any other quadratic to give $[H^+] = 4.63 \times 10^{-10}M$, or pH = 9.33.

This solution is in fact correct to three significant figures, but suppose we had made an excessively simple assumption at the beginning: for example, if we had ignored G^{1-} as well as G^{1+} and G^0 we would have obtained $[H^+] = 7.59 \times 10^{-10}M$ (pH = 9.12), a value that is appreciably in error: how would we have recognized this? Whenever one applies an approximation that one is unsure of it is useful to check its consistency by substituting back into the original expressions. In the glutathione example we can calculate the concentrations of all the species. With $[H^+] = 7.59 \times 10^{-10}M$ we would obtain

$$[G^{3-}] = 0.04M$$

$$[G^{2-}] = 0.04M$$

$$[G^{1-}] = 0.0139M$$

$$[G^0] = 3.57 \times 10^{-8}M$$

$$[G^{1+}] = 3.57 \times 10^{-15}M$$

Although the last two of these are indeed negligible, the value for G^{1-} certainly is not. This alone should indicate that it was wrong to ignore it, but in addition we can calculate a total glutathione concentration of 0.0934M, which exceeds the correct value of 0.08M by a sufficient margin to indicate that the initial approximation was not acceptable.

5.9 Problems

(5.1) Rearrange the following equations so that they express x in terms of y:

(a) $y = x + 3$ (b) $y = 2x^2 + 5$

(c) $y = \ln 5x$ (d) $y = 8 \exp(3x)$

(e) $y = \dfrac{x+4}{x-1}$ (f) $xy + 4 = 3x$

(g) $y = x^2 + 2x - 4$ (h) $\dfrac{y}{x+1} = \dfrac{2}{x-1}$

(i) $\dfrac{y}{3x+1} - \dfrac{1}{x+1} = 1$

(5.2) A metabolite B is produced in one reaction at a constant rate v_1 and is consumed in a second reaction at a rate $V_2[\text{B}]/(K_{m2} + [\text{B}])$. Assuming that a steady state is achieved in which $[\text{B}]$ does not change, obtain an expression for $[\text{B}]$. How large would v_1 have to be for it to become impossible to establish a steady state?

(5.3) Identify any of the following sets of simultaneous equations that are singular and solve the remainder:

(a) $2x + y = -1$
$x + y = -2$

(b) $x + 6y = 3.5$
$12y + 2x = 7$

(c) $8.32x + 1.03y + 39.85 = 0$
$7.18y - 2.22x = 96.14$

(d) $2.34x + 0.63y = 4.31$
$5.03x + 1.35y = 9.27$

(e) $px + qy = r$
$Px + Qy = R$

(f) $n\hat{a} + \hat{b}\Sigma x = \Sigma y$
$\hat{a}\Sigma x + \hat{b}\Sigma x^2 = \Sigma xy$

(Note: in problem (5.3)f treat $\hat{a}$ and $\hat{b}$ as the unknowns and n, Σx, Σx^2, Σy and Σxy as known quantities.)

(5.4) (a) Solve the pair of equations in problem (5.3)f after substituting $x_1 = 1$, $y_1 = 1.27$; $x_2 = 2$, $y_2 = 2.47$; $x_3 = 3$, $y_3 = 3.62$; $x_4 = 4$, $y_4 = 5.08$; $n = 4$. Interpret Σx as the sum of x_i for $i = 1$ to n, and the other summations similarly. (b) What would be the result of the calculation if we put $n = 1$ and used only x_1 and y_1?

(5.5) Evaluate the following determinants:

(a) $\begin{vmatrix} 1 & 2 \\ 3 & 4 \end{vmatrix}$; (b) $\begin{vmatrix} 4.71 & 1.28 \\ 6.43 & 5.11 \end{vmatrix}$; (c) $\begin{vmatrix} -1 & -2 \\ 3 & -7 \end{vmatrix}$; (d) $\begin{vmatrix} 4 & 8 \\ 1 & 2 \end{vmatrix}$

(e) $\begin{vmatrix} 0 & 0 & 0 \\ 1 & 3 & 6 \\ 4 & 2 & 1 \end{vmatrix}$; (f) $\begin{vmatrix} 2.31 & 1.18 \\ -2.22 & 0.47 \end{vmatrix}$; (g) $\begin{vmatrix} 1.7 & 6.4 & 3.1 \\ 2.3 & 8.1 & 1.2 \\ 1.7 & 6.4 & 3.1 \end{vmatrix}$

(5.6) Without carrying out a complete solution, calculate the discriminant of each of the following equations and determine whether there are (1) two unequal real roots, (1a) two unequal rational roots, (2) two equal roots, or (3) no real roots:

(a) $x^2 + 3x + 5 = 0$ (b) $x^2 = 4x - 7$

(c) $2 + 3x = 4x^2$ (d) $x^2 = 2x + 12$

(e) $3x(x + 4) = x + 6$ (f) $\dfrac{2}{x+1} + \dfrac{3}{x+2} = 7$

(5.7) The equilibrium constant for the reaction catalysed by glucokinase has the following value at pH 6.5:

$$\frac{[\text{glucose 6-phosphate}][\text{ADP}]}{[\text{glucose}][\text{ATP}]} = 230$$

If 5mM-glucose is mixed with 4.5mM-ATP and allowed to equilibrate at pH 6.5 in the presence of glucokinase, what is the final concentration of ADP?

(5.8) A mixture of two biochemicals B and C has an absorbance in a 1 cm cuvette of 0.63 at 460 nm and of 0.52 at 500 nm. The molar absorbances, in $M^{-1} cm^{-1}$, of B and C are 1.03×10^3 and 4.57×10^3 respectively at 460 nm and 7.12×10^3 and 1.43×10^3, respectively, at 500 nm. Given that the absorbances of mixtures are additive, and that the absorbance of a solution with molar absorbance A, concentration c and pathlength d is Acd, calculate the concentrations of B and C.

(5.9) Use a graphical method to obtain three approximate solutions to the the equation $x^2 + 7x = 4/(3x + 2)$. Then use Newton's method to refine these to values correct to three places of decimals.

(5.10) Estimate the pH of a 0.1M solution of ammonium lactate, assuming that the pK_a of lactic acid is 3.86, the pK_a of the ammonium ion is 9.26, and $pK_w = 14.0$.

(5.11) Estimate the pH of the solution obtained by mixing 3 ml 0.1M citric acid with 7 ml 0.1M NaOH. For citric acid, $pK_1 = 3.08$, $pK_2 = 4.74$, $pK_3 = 5.40$.

(5.12) For quadratic equations that do not have simple solutions it is sometimes convenient to consider the sum and product of the two roots instead, which are usually simpler. (a) Show this by deriving expressions for the sum and product of the roots of the general quadratic equation $ax^2 + bx + c = 0$; and hence (b) write down the sum and product of the roots of the equation $x^2 - 7x + 5 = 0$.

6 Partial Differentiation

6.1 Meaning of a partial derivative

In science we often have to deal with two or more variables that have no necessary dependence on one another. For example, although Boyle's law defines the following relationship between the pressure p, volume V and temperature T of a mole of a perfect gas:

$$pV = RT$$

(where $R = 8.31 \, \text{J} \, \text{mol}^{-1} \, \text{K}^{-1}$ is the gas constant), it does not prevent us from varying any two of the three variables independently. We can have whatever pressure we like at whatever temperature we like, provided we accept whatever volume results, etc. In this sort of circumstance we cannot define an ordinary derivative, such as dV/dT, because this has no definite value unless we specify how p is to change with T. For example, if we decided that p was to be a constant independent of T we would have

$$\frac{dV}{dT} = \frac{R}{p}$$

but if we set up the system so that p increased in proportion to T, we would instead find

$$\frac{dV}{dT} = 0$$

and other expressions would be given by other relationships between p and T. This is clearly unsatisfactory, but we can overcome the difficulty by introducing the new concept of *partial differentiation*. In this we define the *partial derivative* of one variable with respect to another as the result of differentiating while treating all other

117

independent variables as constants. So, for Boyle's law we would have:

$$\left(\frac{\partial V}{\partial T}\right)_p = \frac{R}{p} = \frac{V}{T}$$

$$\left(\frac{\partial V}{\partial p}\right)_T = -\frac{RT}{p^2} = -\frac{V}{p}$$

$$\left(\frac{\partial p}{\partial T}\right)_V = \frac{R}{V} = \frac{p}{T}$$

There are two points of symbolism to note here. First, as these are not ordinary derivatives, we cannot use the ordinary symbols dV/dT etc. for them. Instead of the ordinary d we use a so-called 'curly d', written as ∂, for partial differentiation. (This is *not* a Greek delta, incidentally, which would be δ and is used in mathematics to represent a small but finite increment, as in Chapter 3.) Second, we can indicate what variables are being held constant during the partial differentiation by showing them as subscripts, as in the expressions above. These are often obvious, however, and may be omitted when no doubt is likely. In thermodynamics they are frequently included as an aid to clarity. In speech we call an expression such as $\partial V/\partial T$ 'partial dV by partial dT', although the second 'partial' is often omitted.

6.2 Exact and inexact differentials

In the previous section we saw that we cannot differentiate one variable with respect to another if the relationship between them is incompletely specified. The same applies to the reverse process, integration, and it is convenient to illustrate it with the same example of a perfect gas because of its great importance in thermodynamics.

Recalling that a *pressure* is a *force* divided by an *area* (or *length* squared), and *work* represents the product of a *force* and a *length*, we can easily see that the work dW done (against the atmosphere, or a piston, or whatever) by a mole of perfect gas at pressure p expanding by an infinitesimal increment dV must be

$$dW = p\,dV$$

We might suppose, therefore, that we could calculate the work W

done in a finite expansion from $V = V_1$ to $V = V_2$ as

$$W = \int_{V_1}^{V_2} p \, \mathrm{d}V$$

However, just as in the previous section we were unable to evaluate $\mathrm{d}V/\mathrm{d}T$ without knowing how V depended on T, here we cannot evaluate $p \, \mathrm{d}V$ without knowing how p depends on V. Further, because Boyle's law contains T as well as p and V, we must specify something about T before we can properly express p in terms of V. In this example $p \, \mathrm{d}V$ is an *inexact differential* and W can in fact have any value.

To integrate $p \, \mathrm{d}V$ we must first define the system fully. Suppose we decide to consider an *isothermal* expansion, which in mathematical terms is the same as defining T as a constant. We then have

$$W = \int_{V_1}^{V_2} p \, \mathrm{d}V = RT \int_{V_1}^{V_2} \frac{\mathrm{d}V}{V} = RT \left[\ln V \right]_{V_1}^{V_2} = RT \ln (V_2/V_1)$$

It is important to realize that we were only able to do this integration, with T outside the integration sign, by specifying T as a constant: if T varied during the expansion the integration would be invalid. Some apparently similar functions turn out to have defined meanings regardless of the way in which the particular change is carried out; these are called *exact differentials*. For example, suppose we had set out to integrate not $p \, \mathrm{d}V$ but $(p/T) \, \mathrm{d}V$. Although at first sight this seems to be just as incomplete as $p \, \mathrm{d}V$, in fact it is a function of V only, because $p/T = R/V$, i.e.

$$\frac{p \, \mathrm{d}V}{T} = \frac{R \, \mathrm{d}V}{V}$$

The factor $1/T$, which converts the inexact differential $p \, \mathrm{d}V$ into the exact differential $(p/T) \, \mathrm{d}V$, is called an *integrating factor* for the inexact differential $p \, \mathrm{d}V$.

Sometimes we find that although two differentials may be inexact, a simple function of them, such as their sum or difference, may be exact. For example, if no relationship between u and v is specified both $\mathrm{d}u/v$ and $u \, \mathrm{d}v/v^2$ are inexact differentials that cannot be integrated. Their difference, however, can be written as

$$\frac{\mathrm{d}u}{v} - \frac{u \, \mathrm{d}v}{v^2} = \frac{v \, \mathrm{d}u - u \, \mathrm{d}v}{v^2}$$

which can immediately be recognized as $d(u/v)$, an exact differential, if we recall the expression for the derivative of a ratio (see p. 59).

This kind of relationship is of the greatest importance in thermodynamics. It is possible to convert any system (not just a sample of perfect gas) from one state to another by adding an indefinite amount of heat $\int dQ$ and by causing it to do an indefinite amount of work $\int dW$. These amounts of heat and work are indefinite because knowing the initial and final states does not allow us to calculate them: their values depend on the path taken between the two states. Nonetheless, it is a fact of observation that although dQ and dW are inexact differentials, their difference $dQ - dW$ is an exact differential. What this means is that although we may observe all kinds of values for the total heat $\int dQ$ and the total work $\int dW$ on passing from state 1 to state 2, we always observe exactly the same difference between them. This is clearly an important observation and is an expression of the *first law of thermodynamics*. The observation that $dQ - dW$ is an exact differential provides the basis for defining the *energy U* of a system, i.e. we define

$$dU = dQ - dW$$

6.3 Least-squares fitting of the Michaelis–Menten equation

Although thermodynamics provides the main contexts in which students of chemistry and biochemistry encounter partial differentiation, it is a rather abstract subject that everyone finds difficult. It is useful therefore to consider a quite different example that has no obvious relationship to thermodynamics but has nonetheless a clear relevance to the practice of biochemistry.

Suppose we have a set of observations (s_i, v_i), for $i = 1, 2, 3 \ldots n$, that fit the Michaelis–Menten equation apart from experimental error, i.e. suppose we can write

$$v_i = \frac{Vs_i(1 + e_i)}{K_m + s_i}$$

where V and K_m are (unknown) constants and $(1 + e_i)$ is a factor representing the effect of experimental error. (We could also have included experimental error by *adding* an error term to the basic expression rather than by multiplying by an error factor. As my purpose in using this illustration is mathematical rather than biochemical it is perhaps unnecessary to discuss why I think the factor is preferable other than that it leads to much simpler mathematics. However, it happens that it is also a more realistic way of representing the kind of error that commonly occurs in enzyme kinetic experiments.) In practice we would not know V and K_m and would therefore wish to *estimate* them from the observed values of s_i and v_i; we might prefer to *measure* them instead, but the unknown magnitudes of the error terms prevent this.

By rearranging the above equation we can readily express each e_i as follows:

$$e_i = \frac{K_m v_i + v_i s_i - V s_i}{V s_i}$$

$$= (K_m/V)(v_i/s_i) + (v_i/V) - 1$$

$$= a x_i + b v_i - 1$$

where $a = K_m/V$, $b = 1/V$, $x_i = v_i/s_i$. (These substitutions are made only to make the mathematical derivation simpler and clearer; they are not essential, inasmuch as we could arrive at exactly the same final result if we carried out the calculation with K_m and V as variables rather than a and b.) We can define the 'best' values of a and b as those that make some suitable function of the e_i as small as possible. First, however, we must choose a function. It is not sufficient simply to add all the e_i together, because some are positive and some are negative, so their sum can be small or zero without the individual e_i having to be small. So we commonly take the *sum of squares S*, defined as

$$S = \sum_{i=1}^{n} e_i^2 = \sum_{i=1}^{n} (a x_i + b v_i - 1)^2$$

as a measure of closeness of fit.

Our aim is now to find values of a and b that make S a minimum. If there were only one parameter this would be a simple problem in differentiation of the sort we considered in Chapter 3. However, we have to find a minimum in S not only with respect to a, but also simultaneously with respect to b. As we can vary a and b independently – there is no relationship that specifies b if we know a, or vice versa – this is an exercise in *partial* differentiation, which we must do with respect to a and b in turn. Differentiating with respect to *a* we treat *b* as a constant:

$$\frac{\partial S}{\partial a} = \sum [2x_i (ax_i + bv_i - 1)] = 2a\Sigma x_i^2 + 2b\Sigma x_i v_i - 2\Sigma x_i$$

Two points should be noted about these expressions: first, as all of the summations are from $i = 1$ to n we can omit the limits from the summation signs without risk of ambiguity; second, in the first and second summations on the right-hand side *a* and *b* (respectively) are factors of every term and so can be multiplied once after summing rather than *n* times before. We can also partially differentiate with respect to *b* in exactly the same way, treating *a* now as a constant:

$$\frac{\partial S}{\partial b} = \sum [2v_i (ax_i + bv_i - 1)] = 2a\Sigma x_i v_i + 2b\Sigma v_i^2 - 2\Sigma v_i$$

For any value of *b*, we can minimize with respect to *a* by finding a value of *a* that makes the first expression zero; conversely, for any value of *a* we can minimize with respect to *b* by finding a value of *b* that makes the second expression zero. To minimize with respect to both parameters simultaneously we must make both expressions zero simultaneously. Let us define $\hat{a}$ and $\hat{b}$, respectively, as the values of *a* and *b* that satisfy this condition, i.e.

$$2\hat{a} \Sigma x_i^2 + 2\hat{b} \Sigma x_i v_i - 2\Sigma x_i = 0$$

$$2\hat{a} \Sigma x_i v_i + 2\hat{b} \Sigma v_i^2 - 2\Sigma v_i = 0$$

These are now a pair of simultaneous equations in $\hat{a}$ and $\hat{b}$, exactly analogous to those we considered in Chapter 5 apart from the fact that we now have rather complicated coefficients such as Σx_i^2 instead of the simple constants we had before. We can therefore use the determinant method to write down the solutions without further algebra:

$$\hat{a} = \frac{\Sigma v_i^2 \Sigma x_i - \Sigma x_i v_i \Sigma v_i}{\Sigma x_i^2 \Sigma v_i^2 - (\Sigma x_i v_i)^2}$$

$$\hat{b} = \frac{\Sigma x_i^2 \Sigma v_i - \Sigma x_i v_i \Sigma x_i}{\Sigma v_i^2 \Sigma x_i^2 - (\Sigma x_i v_i)^2}$$

Finally we can revert to the original symbols to give a result with a more obvious biochemical significance:

$$\hat{K}_m/\hat{V} = \frac{\Sigma v_i^2 \Sigma v_i/s_i - \Sigma v_i^2/s_i \Sigma v_i}{\Sigma v_i^2/s_i^2 \Sigma v_i^2 - (\Sigma v_i^2/s_i)^2}$$

$$1/\hat{V} = \frac{\Sigma v_i^2/s_i^2 \Sigma v_i - \Sigma v_i^2/s_i \Sigma v_i/s_i}{\Sigma v_i^2/s_i^2 \Sigma v_i^2 - (\Sigma v_i^2/s_i)^2}$$

and it is a simple matter to rearrange these into expressions for $\hat{V}$ and $\hat{K}_m$:

$$\hat{V} = \frac{\Sigma v_i^2/s_i^2 \Sigma v_i^2 - (\Sigma v_i^2/s_i)^2}{\Sigma v_i^2/s_i^2 \Sigma v_i - \Sigma v_i^2/s_i \Sigma v_i/s_i}$$

$$\hat{K}_m = \frac{\Sigma v_i^2 \Sigma v_i/s_i - \Sigma v_i^2/s_i \Sigma v_i}{\Sigma v_i^2/s_i^2 \Sigma v_i - \Sigma v_i^2/s_i \Sigma v_i/s_i}$$

These are now the 'best' values of V and K_m in the sense that they make the function S that we defined as our criterion of closeness of fit as small as possible. Before leaving this result it is advisable to check it for algebraic errors by the dimensional considerations we saw in Chapter 1. As $\hat{V}$ is a rate, with dimensions of (concentration) (time)$^{-1}$, its expression must also be a rate: the first sum in the numerator contains terms of the form v^2/s^2, with dimensions of (concentration)^2 (time)$^{-2}$ (concentration)$^{-2}$, or (time)$^{-2}$; the second contains terms of the form v^2, with dimensions of (concentration)^2 (time)$^{-2}$; thus the product of the first two sums has dimensions of (concentration)^2 (time)$^{-4}$. In the same way the squared sum in the second half of the numerator also has the dimensions of (concentration)^2 (time)$^{-4}$ and can therefore be legitimately subtracted from the first product. By the same sort of analysis we find that the denominator also contains a dimensionally acceptable subtraction and has dimensions of (concentration) (time)$^{-3}$. Finally, the expression as a whole has the dimensions of the numerator divided by those of the denominator, i.e. (concentration) (time)$^{-1}$, as it must if it is to be a rate. Thus our expression for $\hat{V}$ is dimensionally correct (though this does not guarantee that it does not contain other kinds of error), and in the same way we can show that the expression for $\hat{K}_m$ is a concentration and is also dimensionally correct.

In this discussion I have treated (concentration) as a fundamental dimension: this is not strictly correct, as a concentration is a derived quantity with dimensions of (amount of substance) (length)$^{-3}$.

Properly therefore I ought to have written (amount of substance) (length)$^{-3}$ wherever I wrote (concentration), but this would not have made the discussion any easier to follow and indeed might have encouraged the kinds of mistakes we were setting out to avoid and recognize. For many purposes there is no harm and a definite advantage in discussing dimensions in terms of composite quantities such as concentrations and even rates.

6.4 Problems

(6.1) For the function $z = (x^2 + y^2)^{1/2}$;

(a) determine $\left(\dfrac{\partial z}{\partial x}\right)_y$;

(b) determine $\left(\dfrac{\partial x}{\partial y}\right)_z$;

(c) determine $\left(\dfrac{\partial y}{\partial z}\right)_x$;

(d) show that $\left(\dfrac{\partial z}{\partial x}\right)_y \left(\dfrac{\partial x}{\partial y}\right)_z \left(\dfrac{\partial y}{\partial z}\right)_x = -1$;

(e) show that $\dfrac{\partial}{\partial y}\dfrac{\partial z}{\partial x} = \dfrac{\partial}{\partial x}\dfrac{\partial z}{\partial y}$.

(6.2) For the function $pV = RT$ (where R is a constant), show that

$$\frac{\partial V}{\partial T}\frac{\partial T}{\partial p}\frac{\partial p}{\partial V} = -1 \text{ and that } \frac{\partial}{\partial p}\frac{\partial V}{\partial T} = \frac{\partial}{\partial T}\frac{\partial V}{\partial p}.$$

(6.3) Consider a set of observations (x_i, y_i), for $i = 1$ to n, that can be represented as values from the following straight-line relationship with additive deviations e_i:

$$y_i = a + bx_i + e_i$$

Defining the sum of squares as $S = \Sigma e_i^2$, find the values of $\hat{a}$ and $\hat{b}$ such that S is a minimum when $a = \hat{a}$, $b = \hat{b}$. [Note: For this problem, in contrast to the one considered in the text, it is easier to deal with an additive error than with a multiplicative error $(1 + e_i)$. A discussion of more fundamental reasons than convenience for preferring one treatment over another would be outside the scope of this book.]

(6.4) Show that your solution to problem (6.3) is dimensionally consistent. (Note: Although no dimensions for x_i and y_i have been specified, dimensional analysis can still be used, as all that is necessary is to assume that the dimensions of x_i and y_i are different.)

Notes and Solutions to Problems

(1.1) (a) 28;   (b) 5;   (c) 41;   (d) 180.

(1.4) (a) Incorrect: $[I]/K_i$ is dimensionless and therefore cannot be added to K_m or $[S]$, which are concentrations; (b) consistent; (c) incorrect: the slope must be a concentration not a reciprocal concentration.

(1.5) (a) 14;   (b) 8.857;   (c) 2;   (d) 17;   (e) 72;   (f) 512; (g) 64.

(1.6) (a) 105 000;   (b) 126 000;   (c) 153 000. Notice how the small proportion of tetramer has only a slight effect on the value of M_n, but a very large effect on the value of M_z. This is because M_z weights the heaviest components of the mixture very heavily. The progression $M_n \leqslant M_w \leqslant M_z$ applies not only to this example but in general.

(1.7) (a) 23.9;   (b) 0.6092;   (c) 0.0624 (although each of the original numbers had 5 or 6 significant figures, 3 were lost in subtracting 35.6112 from 35.6579 to give 0.0467);   (d) 9.2.

(1.8) (a) 8;   (b) 4;   (c) 3;   (d) 4;   (e) 4.

(2.1) (a) 27;   (b) 0.0625;   (c) 3;   (d) 0.5;   (e) 1;   (f) 0.001; (g) 1;   (h) -1.63;   (i) 24;   (j) 27;   (k) 2;   (l) 3; (m) 1000.

(2.2) (a) 1.16; (b) 0.923 ($= 1 - 0.077$). More accurate values are 1.1735 and 0.9259, which show that the approximate formula gives results accurate to within 1.2% and 0.31%, respectively.

(2.3) (a) $\ln(1 + x) \simeq x$;   (b) 0.116 (0.1098);   (c) $-0.017 (-0.0171)$; (d) 0.041 (0.0402);   (e) $-0.112 (-0.1188)$. In (b–e), approximate values given by the formula in (a) are followed in parenthesis by values accurate to 4 places of decimals.

(2.4) (a) $1.386 (= 2 \times 0.693)$;   (b) -1.609;   (c) $3.297 (= 3 \times 1.099)$;
   (d) $-0.510 (= 1.099 - 1.609)$;   (e) $2.302 (= 0.693 + 1.609)$;
   (f) 0.5;   (g) 3;   (h) 15   (as $2.708 = 1.099 + 1.609 = \ln 3$
   $+ \ln 5$); (i) 1.4 ($= 7/5$, as $0.337 = \ln 7 - \ln 5$).

(2.5) About 1.000 016:1 in favour of the state of lower energy
   (calculated with $T = 298$ K, i.e. at $25°$ C).

(2.6) Midpoint potential is $+0.100$ V at pH 0, decreases linearly by
   0.06 V per pH unit to about pH 3.3, then curves gently to a
   smaller slope and decreases by 0.03 V per pH unit above about
   pH 5.3.

(3.1) (a) $3x^2$;   (b) $0.5/x^{1/2}$;   (c) $5e^x$;   (d) $5e^{5x}$;   (e) $3/x$;
   (f) $0.5 (x^{-1/2} - x^{-3/2})$;   (g) $x^{-1} - 4x$;   (h) $(x^2 + 2x)e^x$;
   (i) $3/x$;   (j) $0.5(x^2 + 3)^{-1/2}(2x) = x/(x^2 + 3)^{1/2}$;   (k) $(x - 1)$
   $(1) - (x + 1)(1)/(x - 1)^2 = -2/(x - 1)^2$.

(3.2) (e) and (i), because $3 \ln x$ is identical with $\ln (x^3)$.

(3.3) (a) Minimum, $(3, -1.41)$; maximum, $(1, 0)$. (b) Minimum,
   $(4, 12)$; maximum, $(6, 8)$: note that even though there is only one
   minimum and only one maximum the maximum can occur at a
   lower y value than the minimum. What happens in the vicinity
   of $x = 5$?

(3.4) $dv/ds = K_m V/(K_m + s)^2$,   hence $dv/d \ln s = K_m Vs/(K_m + s)^2$,
   hence $\dfrac{d}{ds} \dfrac{dv}{d \ln s} = (K_m - s)K_m V/(K_m + s)^3$, which is zero if
   $s = K_m$. When $s = K_m$ the slope of a plot of v against $\ln s$ is
   $s^2 V/(s + s)^2 = V/4$. As $\ln s = 2.303 \log s$ (for any s), the maximum slope of a plot of v against $\log s$ is greater by a factor of
   2.303, i.e. it is $0.576V$.

(3.5) $dv/ds = K_m V/(K_m + s)^2$.   (a) V/K_m;   (b) $0.25V/K_m$;
   (c) approaches 0.

(3.6) (a) $h = (K_1 K_2)^{1/2}$.

(3.7) $Y = x^2/(1 + x^2)$; $Y' = 2x/(1 + x^2)^2$; $Y'' = (2 - 6x^2)/(1 + x^2)^3$.
   $Y = 0$ at $x = 0$ and approaches $Y = 1$ at large x. Y' is positive at
   all positive x and so Y increases monotonically with x. Y''
   is positive at the origin but decreases as x increases, being zero at
   $x = 1$ and negative when $x > 1$. Thus the curve is sigmoid (S-
   shaped) with a point of inflexion at $x = 0.577 (= 3^{-1/2})$, $Y = 0.25$.

(3.8) 2 (for $h = 2$); in general it would be h.

(3.9) $V = (dp/dt)[1 + K_m/(s_0 - p)]$, which may be rearranged to
   $dp/dt = Vs/(K_m + s)$, i.e. the Michaelis–Menten equation, if we
   define $s = s_0 - p$.

(4.1) (a) $\frac{1}{3}x^3 + \frac{3}{2}x^2 + x + \alpha$; (b) $\frac{1}{2}x^2 + \ln x + \alpha$; (c) $\frac{1}{3}\ln(2+3x) + \alpha$; (d) $-\frac{1}{3}\exp(-3t) + \alpha$.

(4.2) (a) 37.5; (b) 0.693; (c) 2; (d) 0. In (d) the implication of zero area under the curve between $x = -2$ and $x = 2$ may be puzzling. The explanation is that areas below the x-axis count as negative and in this example the area of -4 between $x = -2$ and $x = 0$ exactly cancels with the area of $+4$ between $x = 0$ and $x = 2$.

(4.3) $y = \dfrac{3}{3x+1} - \dfrac{1}{x+2}$, hence $\int y \, dx = \ln(3x+1) - \ln(x+2) + \alpha$.

(4.4) $\dfrac{1}{D^2}[B(C+Dx) + (AD-BC)\ln(C+Dx)] + \alpha$.

(4.5) This problem is the inverse of problem (3.9). Replacing s with $s_0 - p$ and separating variables we have

$$\int \frac{(K_m + s_0 - p)dp}{s_0 - p} = \int V \, dt.$$

The left-hand side can readily be integrated by substituting $A = K_m + s_0, B = -1, x = p, C = s_0, D = -1$ into the solution to problem (4.4). It is, however, much simpler to note that $s_0 - p$ appears in both numerator and denominator, so that we can write

$$\int \frac{K_m dp}{s_0 - p} + \int dp = \int V \, dt$$

hence $-K_m \ln(s_0 - p) + p = Vt + \alpha$. The constant α is readily evaluated by putting $p = 0$ when $t = 0$, so $-K_m \ln s_0 = \alpha$, and finally $Vt = p + K_m \ln[s_0/(s_0 - p)]$.

(4.6) In this case the short cut used in problem (4.5) is not available and we must use the result of problem (4.4), with the same substitutions as above except for $D = (K_m/K_p - 1)$ instead of $D = -1$. The solution is $Vt = (1 - K_m/K_p)p + K_m(1 + s_0/K_p)\ln[s_0/(s_0 - p)]$.

(4.7) $I = I_0 \exp(-kcx)$, or $\ln(I_0/I) = kcx$. As $\ln a = 2.303 \log a$ for any a, it follows that $kcx = 2.303 Acx$, i.e. $A = k/2.303$.

(4.8) After setting the sum of the two expressions to zero and cancelling AD, we have

$$\frac{\omega^2 x(1 - \overline{V}\rho)M_r c}{RT} - \frac{dc}{dx} = 0$$

Separating the two variables c and x and integrating, we have

$$\ln c = \frac{\omega^2 x^2 (1 - \overline{V}\rho) M_r}{2RT} + \text{constant}$$

Thus a plot of $\ln c$ against x^2 should be a straight line with a slope consisting of known quantities apart from M_r.

(4.9) With four strips, (a) 33.37, (b) 18.51. With eight strips, (a) 33.99, (b) 18.59. Notice that even with four strips one can obtain results almost as accurate as those with eight. The results would be less concordant if the curve were more complicated.

(5.1) (a) $x = y - 3$; (b) $x = \pm [(y - 5)/2]^{1/2}$; (c) $x = 0.2 \exp(y)$; (d) $x = \frac{1}{3} \ln(0.125 y)$; (e) $x = (y + 4)/(y - 1)$; (f) $x = 4/(3 - y)$; (g) $x = -1 \pm (5 + y)^{1/2}$; (h) $x = (y + 2)/(y - 2)$; (i) $x = [y - 7 \pm (25 - 2y + y^2)^{1/2}]/6$.

(5.2) $[B] = K_{m2} v_1/(V_2 - v_1)$. This expression gives a negative value of $[B]$ if v_1 exceeds V_2, and shows that a steady state cannot be achieved if the second reaction cannot remove B faster than it is being produced in the first reaction.

(5.3) (a) $x = 1, y = -3$; (b) singular; (c) $x = -6.21; y = 11.47$; (d) singular*; (e) $x = (Qr - qR)/(pQ - Pq)$; $y = (pR - Pr)/(pQ - Pq)$; (f) $\hat{a} = (\Sigma x^2 \Sigma y - \Sigma x \Sigma xy)/[n\Sigma x^2 - (\Sigma x)^2$; $\hat{b} = (n\Sigma xy - \Sigma x \Sigma y)/[n\Sigma x^2 - (\Sigma x)^2]$. *If the coefficients in problem (5.3)d are treated as exact, the equations are not strictly singular and have the solution $x = 2.18$, $y = -1.26$. This solution is very unstable, however, and would, for example, become $x = 2.53$, $y = -2.57$ if the coefficient of y in the first equation were 0.629 instead of 0.63. It is more reasonable to suppose that the coefficients quoted are accurate only to two decimal places and that the equations are singular. Equations that are nearly singular are called *ill-conditioned*.

(5.4) (a) $\hat{a} = -0.035$, $\hat{b} = 1.258$; (b) the equations would become singular: this illustrates the general point that one cannot determine two constants from one observation.

(5.5) (a) -2; (b) 15.84; (c) 13; (d) 0; (e) 0; (f) 3.705; (g) 0.

(5.6) (a) -11: no real roots; (b) -12: no real roots; (c) 41: two real roots; (d) -44: no real roots; (e) 193: two real roots; (f) 60: two real roots. Rational roots – whether equal or not – occur so rarely as the solutions to scientific problems that it is hardly worthwhile considering them as a possibility.

Attempting to solve quadratic equations by factorization is usually therefore a waste of time.

(5.7) 4.37 mM. The second solution of 5.17 mM is mathematically correct but physically meaningless because it implies negative concentrations of glucose and ATP. In this sort of problem there is no advantage in converting concentrations from mM to M; it leads to inconveniently small numbers and greatly increases the likelihood of arithmetical mistakes. In general it is best to work in units that produce numerical values as close to 1.0 as possible.

(5.8) As in problem (5.7) it is advisable to work in mM rather than M. If a 1 M solution of B has absorbance 1.03×10^3 at 460 nm, a 1 mM solution has absorbance 1.03, and the other values convert similarly. Thus we have $0.63 = 1.03[B] + 4.57[C]$; $0.52 = 7.12[B] + 1.43[C]$, with the solution $[B] = 0.0475$ mM, $[C] = 0.127$ mM.

(5.9) $x = 0.211, -0.908$ or -6.970.

(5.10) 6.56

(5.11) There is sufficient alkali to neutralize the first two acidic groups, so that the significant species are $HCit^{2-}$ and Cit^{3-}. Proceeding as for the glutathione example in the text gives pH = 5.10.

(5.12) (a) sum $= -b/a$, product $= c/a$; (b) sum $= 7$, product $= 5$

(6.1) (a) $x/(x^2 + y^2)^{1/2} = x/z$;   (b) $-y/(z^2 - y^2)^{1/2} = -y/x$;
(c) $z/(z^2 - x^2)^{1/2} = z/y$;   (e) both derivatives have the value $-xy(x^2 + y^2)^{-3/2} = -xy/z^3$.

(6.2) $\dfrac{\partial V}{\partial T} = R/p = V/T, \dfrac{\partial T}{\partial p} = V/R = T/p, \dfrac{\partial p}{\partial V} = -RT/V^2 = -p/V$,

hence the product of the three $= -1$;

$$\frac{\partial}{\partial p}\frac{\partial V}{\partial T} = \frac{\partial}{\partial T}\frac{\partial V}{\partial p} = -R/p^2 = -V/(pT).$$ These relationships, like the similar ones proved in problem (6.1), are not specific to these examples but apply in general.

(6.3) $S = \Sigma(y_i - a - bx_i)^2$;   hence $\dfrac{\partial S}{\partial a} = -2\Sigma y_i + 2an + 2b\Sigma x_i$,

$\dfrac{\partial S}{\partial b} = -2\Sigma x_i y_i + 2a\Sigma x_i + 2b\Sigma x_i^2$ (note that $\Sigma 1$ can be written as n, because it means the sum of n 1's); setting these to zero for $a = \hat{a}, b = \hat{b}$, we have the pair of equations considered previously in problem (5.3)f, which have the solution $\hat{a} = (\Sigma x_i^2 \Sigma y_i - \Sigma x_i \Sigma x_i y_i)/[n \Sigma x_i^2 - (\Sigma x_i)^2]$, $\hat{b} = (n \Sigma x_i y_i - \Sigma x_i \Sigma y_i)/[n \Sigma x_i^2 - (\Sigma x_i)^2]$.

(6.4) In the expression for $\hat{a}$, the first sum in the numerator has the dimensions of x_i^2, the second has the dimensions of y_i, so their product has the dimensions of $x_i^2 y_i$. The second sum also has the dimensions of $x_i^2 y_i$ and can therefore be properly subtracted from the first, to give a complete numerator with the dimensions of $x_i^2 y_i$. In the same way we can show that the denominator also contains a proper subtraction and that its dimensions are those of x_i^2 (in reaching this conclusion we should note that n is dimensionless). Thus the whole expression for $\hat{a}$ has the dimensions of $x_i^2 y_i / x_i^2 = y_i$, which is correct for the intercept on the y_i axis. In the same way we can show that the expression for $\hat{b}$ contains no improper subtractions and that it yield dimensions of y_i/x_i, which is correct for the slope of a plot of y_i against x_i.

Index